AF683518

First published 2025
by CRC Press
4 Park Square, Milton Park, Abingdon, Oxon, OX14 4RN

and by CRC Press
2385 NW Executive Center Drive, Suite 320, Boca Raton FL 33431

CRC Press is an imprint of Informa UK Limited

© 2025 Narendra Publishing House

The right of Pardeep Kumar, Parveen Sharma, Vandana Thakur, Binny Vats and Sunil Kumar to be identified as authors of this work has been asserted in accordance with sections 77 and 78 of the Copyright, Designs and Patents Act 1988.

All rights reserved. No part of this book may be reprinted or reproduced or utilised in any form or by any electronic, mechanical, or other means, now known or hereafter invented, including photocopying and recording, or in any information storage or retrieval system, without permission in writing from the publishers.

For permission to photocopy or use material electronically from this work, access www.copyright.com or contact the Copyright Clearance Center, Inc. (CCC), 222 Rosewood Drive, Danvers, MA 01923, 978-750-8400. For works that are not available on CCC please contact mpkbookspermissions@tandf.co.uk

Trademark notice: Product or corporate names may be trademarks or registered trademarks, and are used only for identification and explanation without intent to infringe.

Print edition not for sale in South Asia (India, Sri Lanka, Nepal, Bangladesh, Pakistan or Bhutan).

British Library Cataloguing-in-Publication Data
A catalogue record for this book is available from the British Library

ISBN13: 9781032873039 (hbk)
ISBN13: 9781032873053 (pbk)
ISBN13: 9781003531951 (ebk)

DOI: 10.1201/9781003531951

Typeset in Adobe Caslon Pro
by Amrit Graphics, Delhi 110032

PRINCIPLES AND TECHNIQUES IN VEGETABLE GRAFTING

First published 2025
by CRC Press
4 Park Square, Milton Park, Abingdon, Oxon, OX14 4RN

and by CRC Press
2385 NW Executive Center Drive, Suite 320, Boca Raton FL 33431

CRC Press is an imprint of Informa UK Limited

© 2025 Narendra Publishing House

The right of Pardeep Kumar, Parveen Sharma, Vandana Thakur, Binny Vats and Sunil Kumar to be identified as authors of this work has been asserted in accordance with sections 77 and 78 of the Copyright, Designs and Patents Act 1988.

All rights reserved. No part of this book may be reprinted or reproduced or utilised in any form or by any electronic, mechanical, or other means, now known or hereafter invented, including photocopying and recording, or in any information storage or retrieval system, without permission in writing from the publishers.

For permission to photocopy or use material electronically from this work, access www.copyright.com or contact the Copyright Clearance Center, Inc. (CCC), 222 Rosewood Drive, Danvers, MA 01923, 978-750-8400. For works that are not available on CCC please contact mpkbookspermissions@tandf.co.uk

Trademark notice: Product or corporate names may be trademarks or registered trademarks, and are used only for identification and explanation without intent to infringe.

Print edition not for sale in South Asia (India, Sri Lanka, Nepal, Bangladesh, Pakistan or Bhutan).

British Library Cataloguing-in-Publication Data
A catalogue record for this book is available from the British Library

ISBN13: 9781032873039 (hbk)
ISBN13: 9781032873053 (pbk)
ISBN13: 9781003531951 (ebk)

DOI: 10.1201/9781003531951

Typeset in Adobe Caslon Pro
by Amrit Graphics, Delhi 110032

PRINCIPLES AND TECHNIQUES IN VEGETABLE GRAFTING

Prof. Pardeep Kumar
Department of Plant Science
School of Life Sciences and Floriculture
Central University of Himachal Pradesh
Kangra - 176206

Prof. Parveen Sharma
Department of Vegetable Science & Floriculture
CSK Himachal Pradesh Agricultural University
Palampur-176062

Dr. Vandana Thakur
Department of Vegetable Science & Floriculture
CSK Himachal Pradesh Agricultural University
Palampur-176062

Mrs. Binny Vats
Department of Plant Sciences
School of Life Sciences
Central University of Himachal Pradesh
Kangra - 176206

Dr. Sunil Kumar
Department of Animal Sciences
School of Life Sciences
Central University of Himachal Pradesh
Kangra - 176206

NPH

CRC Press
Taylor & Francis Group
Boca Raton London New York

CRC Press is an imprint of the
Taylor & Francis Group, an **informa** business

NARENDRA PUBLISHING HOUSE
DELHI (INDIA)

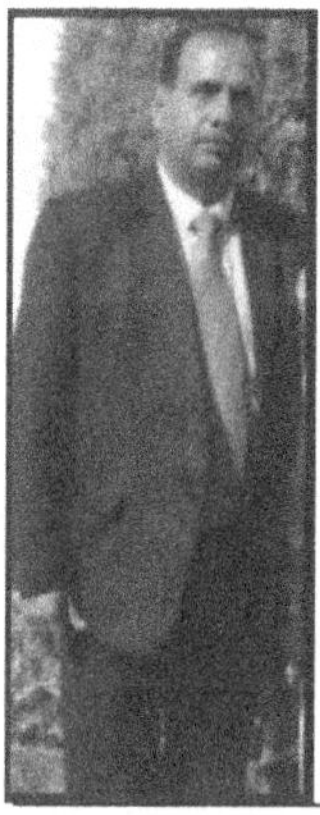

THIS BOOK IS DEDICATED TO
LATE PROF. N.K. PATHANIA, FORMER HEAD,
DEPARTMENT OF VEGETABLE SCIENCE AND
FLORICULTURE, CSKHPKV-PALAMPUR, WHO WAS
MOTIVATIONAL FORCE BEHIND SYSTEMATIC RESEARCH
WORK ON VARIOUS ASPECTS OF VEGETABLE GRAFTING.

CONTENTS

PREFACE

Grafting is an age old concept in plants and is being practiced for centuries all over world. Generally grafting is performed to reduce the vegetative phase in horticultural plants particularly fruit crops. But in vegetables it is a new concept to counter biotic and abiotic stresses by use of resistant rootstocks. India is world's second largest vegetable producer after China. Due to vagaries of weather and outbreak of diseases economic losses to the growers are very huge. Researchers all over world have been continuously working to develop resistant varieties of vegetables, but slow breeding techniques create a big hurdle in managing biotic as well as abiotic factors. There are many modern tools to control diseases and pests such as use of biotechnology, genetic engineering, molecular interventions and new chemistry group of nanotechnology based pesticides or fungicides as well as vegetable grafting.

Research work in vegetable grafting is only 100 years old in some of Asian countries like Japan and Korea, but its very strange that despite so many benefits of this technology it did not become popular. But during last 10 years this technology got recognition from researchers and now many institutions started systematic research work. Some of the private companies also took interest and trying to make it a commercial venture.

This book is a result of 16 years of research experience on various aspects of vegetable grafting. We know that vegetable grafting is a new concept in vegetable crops though it is an age old concept of plant propagation. In horticultural crops particularly in fruits it has been in practice since time immemorial, but in vegetable crops it dates back to 1920 onwards. This book is being written after a lot of practical experiences and discussions with fellow co-workers. We faced a lot of problem in managing bacterial wilt and nematodes under protected conditions as chemical treatments were not effective and resulted in high input costs. Our journey began with a very small experiment and it was so exciting and encouraging that we decided to continue it. We are thankful to state Department of Horticulture and CPRI Shimla for sponsoring a small project under Horticulture Technology Mission (HTM) to initiate a systematic research work on vegetable grafting. The first issue which came across was selection and availability of rootstocks. Rootstocks were imported from AVRDC –Taiwan (World Vegetable Centre) and were multiplied under protected conditions. Compatibility studies were undertaken to find out best rootstock-scion combination. Bacterial wilt and nematode resistant rootstocks proved beneficial in managing the serious problems under protected conditions. We are also thankful to State Department of Agriculture (HP) for providing a financial assistance of Rs. 95 Lakhs under RKVY to undertake vegetable Grafting work on next level. Infrastructure *viz.* Acclimatization chamber, grafting chamber, grafting shed and Grafting Robot was created to facilitate the research as well as extension work.

This book contains comprehensive, current scientific and applied practical knowledge on various aspects of vegetable grafting. The book is based on basic, applied as well as latest concepts in grafting of vegetable crops. The team of experts involved in this work has discussed various practical issues to be faced by the growers and their remedial measures to control the problem. The book is written in a simple and understandable language. Some figures and references were taken from various sources and are fully acknowledged.

In this book **"Principles and Techniques in Vegetable Grafting:"** we will discuss about the rootstocks, their significance, various methods of grafting, use of robotic machines, potential contribution of grafting and future perspectives of this technique toward improvement of vegetable production.

This book is compiled as a collection of scientific information and as a practical tool aimed at both the people involved in the commercial production and cultivation of grafted plants as well as researchers interested in an understanding of the science and technology behind a grafted plant. Special thanks to ***Late Prof. N K Pathania,*** Former Head, Department of Vegetable Science and Floriculture, CSKHPKV Palampur who was the instrument behind this research work. We also express our sincere gratitude's to Prof. S P Sharma, Former Director of Research, CSKHPKV-Palampur for his timely guidance and help during the operations of the research projects.

We also express our sincere thanks to Prof. S P Bansal Hon'ble Vice-Chancellor, Central University of Himachal Pradesh for constant encouragement during writing of this book.

Authors

ABOUT THE AUTHORS

Prof. Pardeep Kumar is presently Head, Department of Plant Sciences and Dean, School of Life Sciences. He started his professional career as Scientist (Vegetable) at Mountain Agricultural Research and Extension Centre-Sangla (Kinnaur) in the year 2006. He has worked as Scientist and Senior Scientist in the Department of Vegetable Science at CSKHPKV-Palampur for 16 years and initiated systematic research work in vegetable grafting. He is also credited with systematic research in development of Pomato and has written more than 50 research papers and more than 100 articles, in various magazines, research bulletins, delivered number of TV and Radio talks. He has delivered more than 1000 lectures to farmers, extension workers and scientists on vegetable grafting, protected cultivation, underutilized vegetable crops and Exotic vegetable cultivation. He has guided one PhD scholar and eight Ph.d. Scholars are pursuing their research under his guidance. He has guided/supervised 25 M.Sc students for their thesis and dissertation work. He has handled 14 research projects as PI/co-PI funded by various funding agencies. He has presented 14 recommendations for the growers of the state. He was awarded State Innovation Award -2016-2017 by the State Government of HP for his innovative vegetative propagation technology for high value vegetables bell-pepper, tomato and cucumber.

Dr. Parveen Sharma started his carrier as an Assistant Scientist at Research Sub Station Akrot (Una) of CSK Himachal Pradesh Krishi Vishvavidyalaya, Palampur in the year 2000 wherein working on protected agriculture; he standardized production technologies of potential vegetable crops for protected environment under different agro - climatic zones of Himachal Pradesh. Presently he is working on breeding vegetables for protected cultivation which resulted in the development of one bacterial wilt resistant hybrid of tomato, one cherry tomato variety rich in beta-carotene and one variety of parthenocarpic cucumber for protected cultivation. Besides, he also developed more than 8 production technologies for protected agriculture and vegetable grafting. He handled 21 external funded ad hoc research projects as Principal Investigator and published more than 78 research papers in reputed journals. In the field of academics, he is involved in teaching of various under-graduate as well as post- graduate courses. Till date he has guided 12 M.Sc. and 4 Ph.D. students as Major Advisor in the field of vegetable breeding. He has also popularized vegetable production technologies under protected environment in dry temperate zone of Himachal Pradesh. He is also involved in the farmer trainings for production of vegetables under protected environment across the country.

Dr. Vandana Thakur, completed Ph.D. in Vegetable Science from Chaudhary Sarwan Kumar Himachal Pradesh Krishi Vishvavidyalaya, Palampur, on Vegetable grafting (Pomato). She was awarded with B.Sc. (Hons.) Horticulture Degree from Dr. Yashwant Singh Parmar University of Horticulture and Forestry, Nauni, Solan (H.P.) and Master's degree from School of Agriculture, Lovely Professional University Phagwara, Jalandhar (Punjab). She was first rank holder in her master's programme. She has served as vocational trainee at Govt. High School Baruna, Nalagrah and worked as Researcher in *My Neta App*, at LPU. She has qualified ASRB-ICAR NET two times. She has published 15 research and review papers in reputed National and International Journals. As an Academic tour, she has visited different parts of India. She is a fellow of Indian Students Association (ISA).

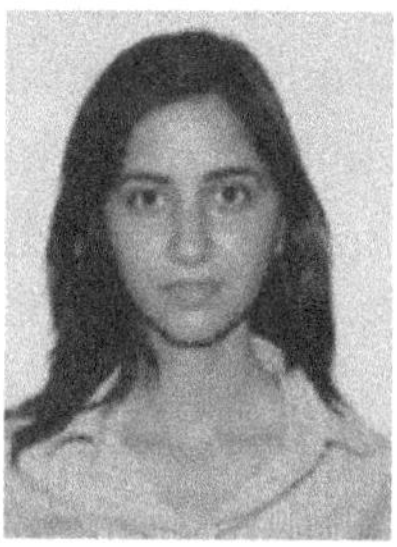

Mrs. Binny Vats, has worked as a Senior Research Fellow in the Department of Vegetable Science and Floriculture, CSKHPKV, Palampur. She is post graduate in Botany with specialization in Plant Biotechnology and has experience in protected cultivation of vegetable crops, grafting techniques, pomato development and vegetative propagation in potential vegetables. She has nine research papers in her credits. Currently, she is pursuing Ph.D. in 'Plant Sciences' under school of Life Sciences, Central University of Himachal Pradesh.

Dr. Sunil Kumar is Associate Professor and Head in the Department of Animal Sciences, School of Life Sciences, Central University of Himachal Pradesh, Dharamshala, Himachal Pradesh, India. He has earned Ph.D. (Entomology and Apiculture) in 2006 from Dr. Y.S. Parmar University of Horticulture and Forestry, Nauni, Solan (H.P.), India. He was awarded University merit scholarship during M.Sc. and Ph.D. After completing his Ph.D. he worked as Assistant Nematologist at CSKHPKV (Palampur) from 2008 to

2010, Assistant Professor at College of Horticulture and Forestry, Neri, Hamirpur (H.P.) from 2010 to 2019 and in September 2019 he joined Central University of Himachal Pradesh as Associate Professor, Department of Animal Sciences. He served as Dean, School of Life Sciences, Central University of Himachal Pradesh from September 2019 to May 2020. He is also serving as Proctor Central University of Himachal Pradesh from March 2021.

Dr. Sunil Kumar has delivered guest lectures in various institutions, more than 15 radio/TV talks on different aspects of entomology, nematology and plant protection to farmers. He has organized various training programs for officials and farmers and delivered lectures in training programs. He has number of publications in his credit and has received financial support for various research projects. He is member of various professional societies.

CHAPTER - 1

GRAFTING: A MAD SCIENCE

People think that it is a miracle of genetic engineering or some rocket science when we talk about pomato/tomtato or brimato two in one plants. Tomato and potato on same plant or brinjal and potato on same plant, now there is one more category i.e. tomato, brinjal and potato on same plant? When we come across grafting innovations like 121 varieties of mango on a single tree or 250 varieties of apple on same plant or a wonder plant known as Tree of Forty (America) where forty types of fruits are available on a single plant then we consider it as a 'Miracle'. **If we think biotechnology was invented in the 21st century, consider the somewhat bizarre practice of sticking the top of one plant onto the bottom of another: Grafting, this is called, and it's been around for at least 4,000 years.** There are many interesting scientific stories where grafting played astonishing role in combating serious problems in horticultural plants.

French Pandemic: Beginning in 1864, and without warning, grapevines across France began to sharply decline. Thanks to the efforts of scientists such as C. V. Riley and J. E. Planchon, the culprit was identified to be phylloxera, an insect that infests the roots of vines and causes fungal infections. Initially, farmers unsuccessfully attempted to contain the pest by removing and burning affected vines. When it was discovered that phylloxera was an invasive species introduced from North America, some suggested importing rootstock from the region as the North American vines were resistant to the pest. Others, opposed to the idea, argued that American rootstocks would imbue the French grapes with an undesirable taste; they instead preferred to inject the soil with expensive pesticides. Ultimately, ***grafting American rootstock onto French vines became prevalent throughout the region, creating new grafting techniques and machines.*** American rootstocks had trouble adapting to the high soil pH value of some regions in France so the final solution to the pandemic was to hybridize the American and French variants. An aphid typically called *phylloxera vitafoliae*, somehow managed to travel from the New to the Old World and swept through Europe in the second half of the 19th century. In its trial it left shrivelled, fruitless vines on desolate vineyards with confused proprietors in a region of the world that relied heavily upon wine. Wine production in France fell 72% in 14 years and put many small, individually owned vineyards out of business. With the combined work of entomologists, biologists, viticulture societies and governments a solution was found. **European varieties could be grafted onto American grape rootstock, which because *phylloxera* is native to the Americas, was virtually immune to its effects.** It's amazing to think that by 1990, about 85 per cent of all the world's vineyards were grafted onto American root stocks that were resistant to *phylloxera.*

Hon'ble Prime Minister of India gave a clarion call to the farm scientists to work out strategies for doubling farmer's income. Many innovative ideas are being implemented to reduce the cost of cultivation, so that by reducing the cultivation expenses income of the farmers may be doubled. Modern technologies are also being used such as hydroponic, aeroponic, vertical farming, protected cultivation, nanotechnology, precision farming to increase the profit margins of the poor farmers. In our opinion from experience of more than 10 years in vegetable grafting, it can also contribute to double the income of farmers. One such initiative is being done by ICRISAT-Hyderabad where more than 400 vegetable producers in India's Andhra Pradesh state have taken to cultivating grafted vegetables in an attempt to double their incomes through increased yields. Farmers are reporting around 30-50% increase in yields from use of grafted varieties over traditional varieties. The farmers, from villages around Kuppam town of Chitoor district, are being provided grafted seedlings as part of an ICRISAT-led project, *'Doubling farmers' income through grafted vegetable seedlings in Andhra Pradesh*, India'. At a Centre of Excellence in Kuppam, from where the project is being coordinated, farmers are trained in the basics of managing grafted vegetables like **tomato, capsicum, chilli, bitter gourd, cucumber and snake gourd**. Mr Chowdappa, a farmer from Ramkuppam village said he took to cultivating bottle gourd (grafted variety).

Dr.Narayan Chawda- A Successful farmer and enreprenuerer did commendable work on Brinjal Grafting at Chhatisgarh. Similarly CSKHPKV in 2010 started working on finding a solution to bacterial wilt and nematode problems under protected conditions. Department of Vegetable Science and Floriculture took initiative and started collection of rootstocks for this purpose. Number of rootstocks was screened under field as well as laboratory conditions for bacterial wilt and nematode reaction. Department started working on grafting innovations and successfully developed pomato and brimato. People visiting pomato fields call it a Miracle, but it is a simple technology which is in use since thousands of years. Many NGOs awarded us for this innovation work and we provided number of practical trainings to the farmers, scientists, agricultural officers and stakeholders to learn this mad science. Grafting is used in various ways to increase fruit production, to reduce vegetative phase, to manage biotic and abiotic stresses as well as for aesthetic purposes. In vegetables it is a relatively new concept and farmers initially were hesitant and reluctant to learn this technology. But, now numbers of institutions are after this mad science to create new recombinant plants as well as to provide disease resistance to the plants. We have tried to discuss various issues of vegetable grafting in a very simple manner so that even a layman can understand the concept and adopt it for his benefit.

Farmers learning grafting techniques in vegetables

CHAPTER - 2

HISTORICAL BACKGROUND OF VEGETABLE GRAFTING

2.1 HISTORY OF VEGETABLE GRAFTING

Vegetable grafting is an exceptional horticultural technique practised, since many years in East Asia to cope the obstacles allied with intensive cultivation of vegetable crops. It's a centuries-old-technique but more likely to be new in the area of vegetable production. A historic book written in China at the time of fifth century and in Korea in seventeenth century described self-grafting technique which was used to harvest a large gourd fruit by escalating root-to-shoot ratio through more than a few grafting (Lee and Oda, 2003). The first reference of interspecific, vegetable grafting to increase yield and control pest/disease was used for watermelon [*Citrullus lanatus* (Thumb.) Matsum. & Nakai], grafted on squash rootstock (*Cucurbita moschata* Duch), purportedly developed by a water-melon growing farmer in Japan (Tateishi, 1927). Meanwhile, this watermelon grafting approach was popularized in the midst of farmers through extension research programmes started by regional agricultural experimental stations in Japan in the year 1920 and then later into Korea in early 1930s.

Watermelon grafted on *Lagenaria siceraria* (Mol.) Standl. was promoted for commercial use in Japan during 1930's (Oda, 2002). Meanwhile the research trials on cucumber grafting (*Cucumis sativus* L.) was ongoing in the late 1920s, but commercial use was started in 1960 (Sakata et al., 2008). In the list of potential vegetables, particularly belongs to solanaceous family utilization of grafting approach was first time recorded in eggplant (*Solanum integrifolium* Poir.) in the year 1950s (Oda, 1999). Grafting in tomato (*Solanum lycopersicum* Mill.) was commercially launched in the year 1960s (Lee and Oda, 2003) with the speedy expansion of protected cultivation technologies (high tunnels and greenhouses). Vegetable grafting became a significant means to manage soil borne diseases and other pests. At some point in 1990s, just about sixty per cent of open fields and greenhouses in Japan performed the grafting in cucumber, muskmelon, watermelon, tomato and eggplant (National Research Institute of Vegetables, Ornamental plants and Tea, 2001); while in Korea about eighty one per cent of grafted cucurbits, tomato and egg plants were grown (1990 data reported by Lee, 1994).

Intensive labour input ensuing high cost of grafted seedling production is becoming the foremost key obstacle in the popularity and from being widely adopted outside the Asia. Due to sincere efforts of multinational seed companies and researchers working on vegetable grafting along with improved production techniques for grafted seedlings and new compatible rootstocks, this technology was introduced in European countries during early 1990's (Oda, 2002). The foremost objectives of using grafted seedlings are: firstly to get resistance from numerous soil borne diseases and nematodes, secondly to augment yield and quality and thirdly to improvise the physiology of plants making them further adaptable to harsh environments. Thus, several European countries, the Middle East, Central America, Northern Africa and other parts of Asia (apart from Japan and Korea) adopted the technology and area under grafted plants augmented promptly throughout the past two decades. Until in recent times, grafted seedling production and its exploitation were not common in North America. Tomato grafting was practiced at a some degree of scale in the US more than sixty years ago. Tomato grafted onto jimsen weed (*Datura stramonium* L.) rootstock was practiced for many years in the southern US to overcome root-knot nematodes as per the confirmation with findings of Lowman and Kelley (1946).

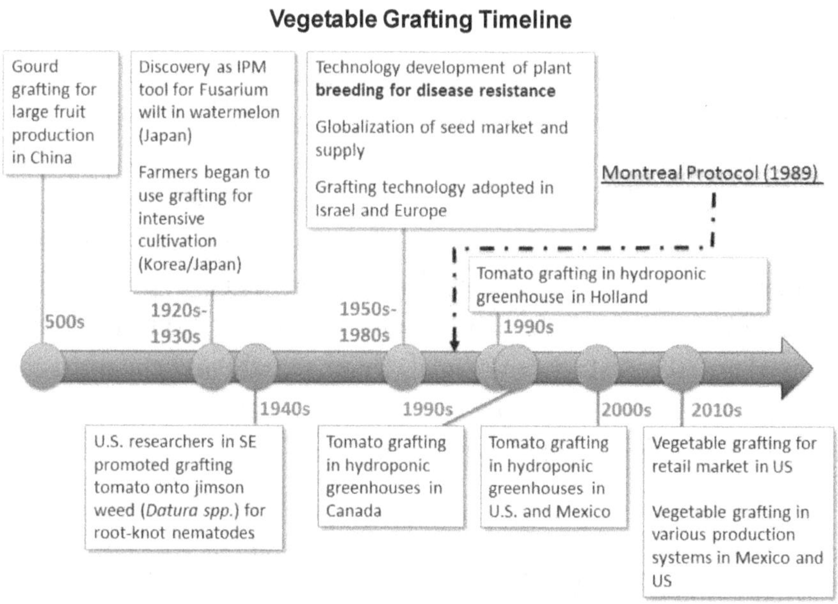

Fig. 1: Timeline showing History of vegetable grafting in the world (upper part, blue) and in North America (lower part, red).
(**Source:** Chieri Kubota The Ohio State University)

According to the research reports confirmed by Isbell (1994) in the year 1935 and 1943 grafting technique was performed in tomato, eggplant and sweet pepper on selected weed rootstock. Nevertheless the combination of tomato (scion) and jimson weed (rootstock) was not introduced on commercial scale and seems to be disappeared, due to the transportation of small amount of alkaloids to fruit of the tomato and labour intensive propagation method has been confirmed by Lowman and Kelly (1946). In 2013, grafted "Pomato" plant was introduced in UK by a private seed company 'Thompson and Morgan' later they launched another wonder plant known as Egg and chips plant in 2015 (Kumar et al., 2015).

History in India

IIHR Bengaluru initiated research for identification of rootstocks for waterlogged conditions. Whereas NBPGR regional station, Thrissur, Kerala have done work on Cucurbit by grafting female plants successfully on the male plants with 98% graft success in *Momordica cochinchinensis*, a dioecious plant. Chaudhary Sarwan Kumar Himachal Pradesh Krishi Vishvavidalaya, Palampur initiated systematic research work on vegetable grafting in 2011 and found more than 30 rootstocks of brinjal, chilli, tomato and cucurbits for imparting resistance to serious diseases like bacterial wilt, nematodes and Gummy Stem Blight (*Didymella broyanae*).

Likewise successful work has been done to grow 'Pomato' double crop plant and already conducted experimental trials to popularize Pomato since 2015. Department of Vegetable Science and Floriculture, CSKHPKV-Palampur is the first agricultural university in the country to install semiautomatic 'Grafting Robot' in July 2017 to upscale grafted vegetable research work. Recently, TNAU-Tamil Nadu has also purchased and installed semiautomatic grafting machine to improvise the grafting work especially with focus on cucurbits.

Navsari Agricultural University, Gujarat has also initiated research work on vegetable grafting. There are many other agricultural research institutions in the country that have recently started work on vegetable grafting. In India, some private companies are also showing their interest in this new era technology to explore this vision throughout the country among farmers. 'VNR Seed Private Limited' in Chhattisgarh is one of the private seed companies, which is supplying bacterial wilt resistant grafted brinjal seedlings to farmers.

2.2 HISTORY OF ROBOTIC GRAFTING

The very first modernized grafting robot "one-cotyledon splice grafting" was developed in Japan by Iam Brain in 1980's, particularly for performing grafting operation in cucurbits. The seedlings are cut from the attachment site of the cotyledon to the hypocotyl at the angle of 20-30° for both rootstock and scion, respectively. In addition to this, the "prototypes grafting robot" was manufactured in the year of 1987 and the second one in 1989; but the prototype semi automated grafting system was developed by Korea based private company with reasonable price, multiple functioning facility and ease to handle. The main advantage of robotic grafting is that it takes on an about 4.5 seconds with 95% survivability. On an accounting an another achievement was achieved for sorting the tomato seedling algorithm with 97% success rate for a full automatic grafting machine as mentioned by Ashraf et al., (2011). A rotator grafting machine with 88% of success rate has been developed in China (Zhou et al., 2009).

In Taiwan, Lee et al. (2001) created automatic grafting robot for passion melon which was able to carry out grafting of 114 seedlings per hour with success rate of 70% and survival success of 95 %. Chen et al. (2010) reported that an automatic tubing grafting robotic system was developed with facility of grafting 327 seedlings per hour with success of 93.8 percent.

Chiu et al. (2010) evolved an automatic robot for grafting watermelon seedlings in Taiwan. This system developed a top plug in grafting robotic machine /system mainly used in grafting of a scion on a matured rootstock of a plant, but before the spread of scion cotyledon. The important point that has to be taken into consideration is that the scion should be very delicate and differs in terms of seedling age from rootstock. This system is characterized by the absence of grafting clips and performed all the grafting procedure and operation accurately; and found to be very successive system with average success rate of about 95 per cent with working capability of producing four hundred eighty grafted seedlings per hour.

Japan is one of the countries in worldwide actively engaged in development of grafting system, where the use of grafting technique has been performed from a very long period of time very frequently. Various types of grafting robots have been developed by the ISEKI Company. These grafting robots are modified from the ones developed by the Biology Research Organization and then commercialized. The working capacity of these robots reaches about 900 seedlings per hour with three persons operating them and has a success rate of more than 95%. This type of grafting robot employs clips to fix the grafting junction of rootstock and scion and it is most suitable for rootstocks and scions that are about the same age. The Bio-oriented Technology Research Advancement

Institute developed a grafting robotic system for cucumber, with capability of producing about eight hundred fifteen grafted seedlings per hours and the success rate counted about 97.1 per cent as about three times for grafts produced than manual grafting.

2.3 CONCLUSION

Vegetable grafting has been practiced for many years in Asia, Europe, the Middle East, Northern Africa, Central America and now it has been introduced as a relatively new technology in North America. As a result of its benefits and value, demand for high-quality grafted seedlings by growers and interest by propagators are expected to rapidly increase. Researchers, extension specialists and industries need to work in collaboration to popularize this technology.

CHAPTER - 3

VEGETABLE GRAFTING: IMPORTANCE AND SCOPE IN INDIA

3.1 CURRENT SCENARIO OF VEGETABLE PRODUCTION

China is a world leader in vegetable production and is followed by India in global market. On account of achievements of horticulture sectors in recent time it indicates the possibilities and facts that the demand of vegetable production is increasing. According to USDA, National Agricultural Statistics Service, in 2016, the production of India for 26-estimated vegetable and melon crops was 780 million cwt. Whereas, total utilized production for vegetables contributed about 775 million cwt. (centum weight) and area was 2.57 million acres. India shares about 10.6% of total vegetable production of the world.

On the basis of recent data of National Horticulture Board 2017-18, total area and production of horticultural crops are 25431 (000 Ha) and 311714 (000 MT), respectively. The present area under vegetable crops is 10259 (000 ha) with 184394 (000 MT) production as per 2017-17 NHB data.

Promising quality and infrastructure ensures the progress in the area of state-of-the-art cold chain and improvised the post-harvest management throughout the country. Public and private sectors play important role in expansion of horticulture sector by

Table 1: Current Status of Vegetable Production (NHB, 2017-18)

Vegetables	Area ('000 Ha)	Production ('000 MT)
Beans	228	2277
Bitter gourd	97	1137
Bottle gourd	157	2683
Brinjal	730	12801
Cabbage	399	9037
Capsicum	24	326
Carrot	97	1648
Cauliflower	453	8668
Cucumber	82	1260
Chillies	309	3592
Elephant foot yam	30	774
Mushroom	198	487
Okra	509	6095
Onion	1285	23262
Parwal/Pointed gourd	20	310
Peas	540	5422
Potato	2142	51310
Radish	209	3061
Pumpkin	78	1714
Sweet Potato	131	1500
Tapioca	173	4950
Tomato	789	19759
Other	1580	22320
Total Vegetables	**10259**	**184394**

(National Horticulture Board, 2017-18)

introducing high value and low volume products for consumers to ensure the availability of food as well as seeds to farmers. Both sectors have also taken initiatives along with APEDA's assistance to setup several "Centers for Perishable Cargoes and Integrated

Post-Harvest Handling Facilities" throughout the country. All these efforts can encourage the farmers to expand the area in vegetable production which ultimately increases the production and productivity up to a greater extent, where we will be able to ensure to provide quality food to the future generations.

In order to maximize the yield and profitability most of the Indian growers are using heavy doses of synthetic chemicals and fertilizers, affecting the food chain and leads to health and environmental hazards. Other environment friendly options like genetically engineered or modified crops which are effective mechanisms to increase yield and suppress biotic and abiotic stresses, but are not accepted by a large section of consumers due to confusing misconceptions. Under such situations it is desired that other farmer friendly techniques are to be explored and vegetable grafting is one of the available options being quick, practical and eco-friendly in nature. Asian countries using vegetable grafting for many years to manage various vegetable production problems such as disease incidence and abiotic stress tolerance as well as soil and water pollution stress. Due to health consciousness issues people are avoiding inorganically grown crops and have liking for organic and naturally grown food crops and vegetable grafting is becoming a new viable option to reduce dependence on chemicals to counter various stresses.

Results of various research experiments carried out on vegetable grafting of solanaceous and cucurbitaceous crops for up gradation of growth, yield and quality of crops has emphasized the significance of grafting in managing these stresses. Keeping in mind, the importance of grafting throughout the globe, there is an urgent need to focus research work in India to control various soil borne diseases and abiotic stresses.

During 1948-50 government focused on cereals production in the country to feed increasing population. Consolidation of institutional support and a planned process for the development of horticulture was initiated during the period 1980-92. Focused attention was given to horticulture development through an enhancement of plan allocation and knowledge based technology during post-1993 period. To promote horticulture in the country a centrally sponsored scheme "National Horticulture Mission" was launched in April 2005 to accelerate holistic growth in horticulture sector with area based regionally differentiated strategies.

Foreign trade policy in 2004-09 boosted agricultural exports and promoted horticultural products. Asian neighboring countries like Bangladesh, UAE, Pakistan, Malaysia, Sri Lanka, UK, Saudi Arabia and Nepal are nearby and major markets for export of vegetables and fruits.

3.2 OBJECTIVES

Table2: Specific objectives of grafting in different vegetable crops are:

Crop	Objectives
Tomato	Tolerance to bacterial wilt (*Pseudomonas solanacearum*), *Fusarium oxysporum*, *Verticillium dahlia* and nematodes (*Meloidogyne spp.*)
Bell Pepper	Tolerance to bacterial wilt, phytophthora, mosaic viruses and nematodes.
Cucumber	Tolerance to fusarium wilt, low temperature, *Phytophthora melonis*.
Brinjal	Tolerance to bacterial wilt, verticillium, nematodes, greater vigour.
Watermelon	Tolerance to fusarium wilt (*Fusarium oxysporum*), low temperature, wilting due to physiological disorder, drought tolerance.

3.3 IMPORTANCE AND SCOPE IN VEGETABLE PRODUCTION

Scope in Vegetable Production

To achieve higher productivity and yield potential of vegetable crops, implementation of pro-active government policies for the development of technologies, high yielding varieties and law enforcement is necessary, the execution and support for such kind of promotional activities needs:

- Expansion of area under improved or hybrid varieties
- Crop diversification
- Conservation of germplasm
- Protected cultivation/ off season vegetable production
- Improvement of peri- urban vegetable farming
- Integrated pest management
- Precision farming
- Vegetable grafting/ Use of Biotechnology
- Breeding for resistance from biotic and abiotic stress
- Breeding for improvement in nutritional and processing quality
- Establishment of advance plant health clinics
- Use of improved modern techniques for post-harvest management
- Establishment of agri-business consortium/ farmer's cooperative society
- Risk assistance to farmers
- Establishment and strengthening of regular fresh vegetable markets, introduction of grading standards.
- Vocational courses for farmers and entrepreneurs
- Integrated Plant Nutrient Management System (IPNS) strategies
- Front line demonstration

- Increased seed replacement
- Public and private sector tie ups for seed multiplication
- Seed village clusters/ organic farming/ cluster farming/ export zones

(Hazra and Som, 2015)

Importance: Vegetables are the most important part of the daily diet and are the rich source of vitamins, carbohydrates, minerals and other phyto-nutraceuticals compounds. Vegetables do not ensure nutritional security in the population until concept of "3A" is not fulfilled i.e. (1) Availability/ Accessibility, (2) Affordability and (3) Acceptability. Vegetable industries have ample scope and significance in development and growth of horticulture sector, which is mentioned below:

ADVANTAGES OF VEGETABLE CROPS

Nutritional security	Production of more biomass	Reduction in malnutrition	Digestible protein	Economical to grow
Well-fitting in farming systems	Suitable for mixed, companion and intercropping	Maximum output and more income per unit area	Suitable for small farmers	
Source of supplementary income	Intensive year round employment	Export potential		
Higher income				

(Hari Har Ram, 2017)

3.4 CONSTRAINTS IN VEGETABLE PRODUCTION IN INDIA

Environmental stresses are the main factor which affects the plant physiological and metabolic activities, leads to hinder the plant growth and ultimately decreases the rate of production and productivity. These constraints are broadly categorized into two parts i.e. biotic and abiotic stress.

Table 3: Sources of environmental stresses in plants are:

	Abiotic Stress	
Biotic Stress	**Physical**	**Chemical**
Temperature	Soil Ph	Diseases
Drought	Salinity	Pathogenic fungi
Flooding	Toxins	Viruses
Radiation	Pesticides	Allelopathy
Magnetic field	Air pollution	Herbivory
Wind	Heavy metals	Competition

(Nilsen and Orcutt 1996)

Biotic Stresses

Biotic or biological stresses are the key factors which affect the metabolism of the living plant and are responsible for the reduction of yield in agricultural crops. Profitability of agriculture and horticulture crops depend on the total productivity of the crops. As a result these biotic stresses become obstructions aimed at cost management for seasonal cultivation of crops. The degree of incidence of biotic stress is genetically controlled and depends on the level of tolerance/resistance shown by the plant towards pest. Pest, weeds and vertebrates e.g. rodents are the foremost causes for the reduction in crop yield.

However in terms of losses, insect damage is about 25%, due to vertebrates and rodents 6-8% and disease including nematodes crops is lost to the tune of 20% approximately. In high volume commodities, such kind of losses in different geographical areas has become the most challenging hurdle in vegetable production.

Apart from these stresses global population is increasing hastily and likely to reach 10 million by 2050, which will put more pressure on agriculture to fulfill the requirement of food and nutrition for human population in the country. Due to rapid increase in population it is not possible to expand the agricultural area in horizontal, so it's necessary to take it into consideration as to feed the millions of population by utilizing such kind of techniques which helps to utilize vertical space more efficiently and maximize the yield and productivity of the country. So, the foremost priority is to increase crop production from limited land and resources.

Now a day's agricultural researcher's focus is to develop biotic/abiotic stress tolerant/resistant crops. About 33% cultivated land in India is under irrigation, whereas 67%, land is dry, or rain fed, due to limited resources which makes crops prone to biotic stresses. Lack of access to modern tools and products is the major reason for fewer profit margins in agriculture. Agricultural scientists are planning strategies to counter biotic factors responsible for reduction in vegetable production through "Hub and Spoke" network.

There are numerous factors which could be the reasons for increasing pest pressure due to propagation of new variants of pathogens which may be aggressive and detrimental for the growth, yield and ultimately quality of the crop.

Increasing urbanization and industrialization also affect ecosystem by polluting air, ground water by producing inorganic pollutants, heavy metals which affect ground water and manifesting greenhouse gases in the atmosphere.

Use of pesticides and fungicides in uncontrolled limits is another factor which makes pathogens resistant and cause huge economic losses to the growers. To get higher returns per unit area farmers use excessive doses which imbalances the natural ecosystem and food chain, which results in new variants in environment which has no effective chemical control mechanisms, hence reduce the total yield of the crop.

Integrated pest management (IPM) is a key strategy to limit the incidences of aggressive pests. It is a balanced application of all pest suppression tools in the IPM tool box to restore the existing food chain in agricultural farms. Many serious diseases such as potato late blight, bacterial wilt of tomato, shoot and fruit borer of brinjal etc. were restricted due to sincere breeding programmes to develop resistant/tolerant varieties/cultivars.

Latest developments in modern chemical science helped in formulation of new chemistry environment friendly pesticides with less residual effects and targeted pest control management systems controlled major outbreaks of mites, jassids and aphids in many commercial agricultural crops. This new category of pesticides can be therapeutic and may be helpful in restoring natural agro-ecological food networks. Despite all these technologies, still abuse of pesticides in agriculture to manage many serious pests and diseases are uncontrolled. Farmers due to lack of awareness and proper guidance use incorrect application of pesticides and fungicides and this result in lower benefit cost ratios.

Abiotic stresses

Certain environmental factors that influence normal plant development, growth, and productivity, exceed thresholds (species-specific) and damage the plant. Drought, cold (chilling and freezing), salt, heavy metals, heat shock, anoxia and nutrient stress are the major abiotic stresses which causes serious problems in vegetable production.

According to recent estimates, 24.2% of the world's geographical area is arable. Only 10.6% of the geographical area is under cultivation and rest is under abiotic stresses. In India 13.6% area is potentially, but not actually arable. Drought is the main abiotic factor as it affects 26% of the arable area. Mineral toxicities/ deficiencies with 22% are second in importance, while freezing stands close third with 14% area under stress. Quantification of degree of an abiotic stress is generally difficult. The severity of the stress on crop depends greatly on duration and stage of crop growth. Region/location specific importance of abiotic stresses, occurrence and location of stresses is unpredictable. Degree of some stresses varies during cropping season. Some stresses can be managed e.g. drought, salinity whereas some cannot e.g. temperature stress. One abiotic stress may increase or decrease the effect of other stress e.g. high moisture stress can increase salinity stress. Different crop plants/varieties/species responds differently to various stresses, Crop growth stage affect tolerance to various stresses. Stress during reproductive phase cause more stress as compared to vegetative stage. The effects generated by one abiotic stress may overlap some of those generated by another stress.

Biotic and abiotic stresses are the major reasons for reduced productivity of vegetables in the country. Resistant varieties are scanty and there is also non-availability of seeds of resistant varieties. To counter these stresses worldwide now grafting of susceptible scions on resistant rootstocks is recognized as viable and easy technology.

3.5 IMPORTANCE AND SCOPE OF VEGETABLE GRAFTING

Vegetable grafting is becoming popular in the country due to concerted efforts of the researchers and extension workers. There is an urgent need to:

1. Identify resistant/tolerant rootstocks for biotic and abiotic stress
2. Availability of healthy seedlings at reasonable price
3. Development of cost effective techniques for farmers
4. Need to conduct research work to minimize the post grafting losses
5. Ample scope for breeder and private sectors companies to develop resistant rootstocks
6. Introduction of modernized technology and extension specialist are required
7. Sharpening the grafting skills
8. Ensures the availability of standardized healing environment
9. Extending the area under vegetable grafting
10. Conduct extension programmes to popularize the technique among farmers

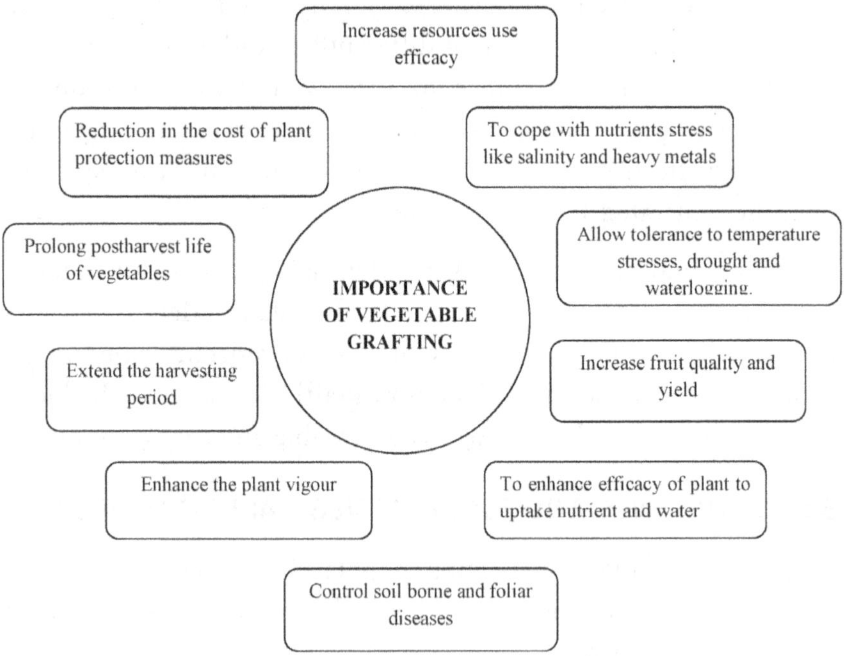

3.6 CONCLUSION

To develop a resistant variety for any kind of environmental stress it will take longer time of about 10-12 years. During this process of developing the resistant varieties with continuous change in climatic condition, number of new race of pathogen develops in environment, and these new race found to be more virulent, which ultimately leads to breakdown of resistance. Another problem associated with resistant varieties is non-availability of the seeds to the farmer. So, to deal with such kind of hurdles and to utilize longer time period in an efficient manner, grafting technique is commercially used in many countries. Grafting of susceptible scion on resistant rootstock is only the alternative way to deal with it. Since, India has major share in vegetables after China and there is a need to feed the population of about 130 million with very limited resources within short period of time. Now a days, many agricultural universities and institutions have also initiated research work in vegetable grafting throughout the country to counter the major environmental stresses in order to maximize the productivity of the country.

CHAPTER - 4

VEGETABLE GRAFTING: CONCEPT, PRODUCTION PROCESS, PRESENT STATUS AND BENEFITS

4.1 INTRODUCTION

Grafting is the spontaneous or intentional union of plant components so that vascular connection develops between them and the resulting plant acts as a single plant.

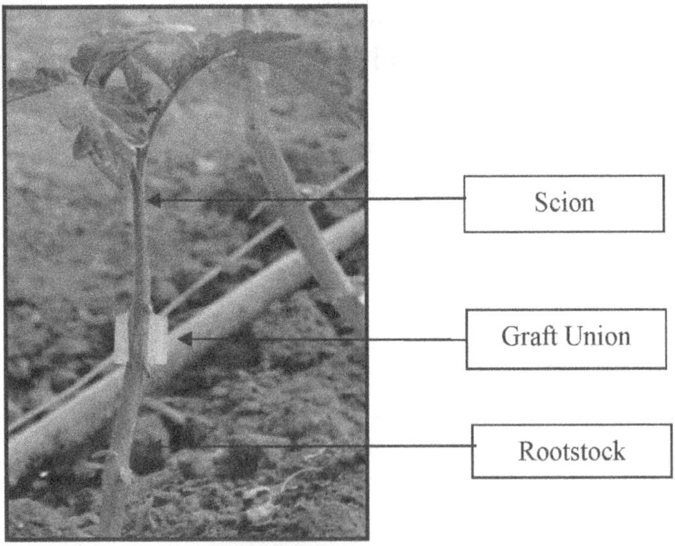

A Grafted Plant

It is a process of joining of living plant tissues to grow and develop as a single plant. In a grafted union two plant parts are **"Rootstock"** which provides the strong and vigorous root system where as **"Scion"** produces the quality fruit.

Grafting in woody plants, horizontal crops and ornamental plants has been common for centuries whereas in vegetables it is needed mainly to develop tolerance/ resistance into plants against stresses. Vegetable grafting is the most promising technique which not only manages biotic stress like soil borne diseases, but also enhances fruit quality; improve crop response to abiotic stresses *viz*, including salinity, moisture, temperature and nutrient extremes etc. It can be used in different ways for improvising the societal, economic and environmental benefits by extending the knowledge about fundamental processes of plant science.

It was devised to avoid serious crop losses caused due to soil-borne pathogens and aggravated by successive cropping. A grafted seedling has been predominant in the production of vegetables such as tomato, pepper, eggplant, cucumber and melon. It is an environment-friendly unconventional technique to manage soil borne pathogens and to maximize the productivity of susceptible cultivars in terms of higher yield.

The success of grafting depends on the formation of vascular connections between stock and scion where hormones from both components initiate cell division in parenchymatous cell in the cut region resulting in callus formation by intermingling and interlocking to form the callus bridge. Hormones specifically auxins and cytokinins play a major role in the early establishment of compatible graft union.

Grafted plant with fruits

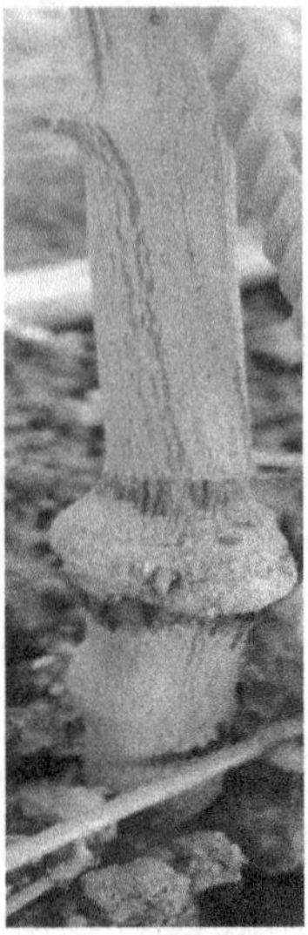

Well established grafted union

Grafting reduces dependence on agrochemicals therefore, considered to be eco-friendly for sustainable vegetable production. Due to the unrelenting loss of quality and disease-free farmland, as well as the phase-out of methyl bromide, vegetable grafting has become an appealing choice, and it is currently quickly spreading and developing all over the world. According to estimations, Japan produces 92% of watermelons, Korea produces 98%, and China produces 20% of watermelons through grafted seedlings. In Europe, Spain leads with 129 million grafted seedlings, followed by Italy with 47 million grafted seedlings and France with 28 million grafted seedlings. The future thrust should be on identification of compatible multiple disease resistant rootstocks with tolerance to biotic and abiotic stresses, techniques of grafting supposed to be of low cost, so they could be adopted by farmers for commercial level production. Meanwhile, the research efforts should be concentrated on minimizing the mortality rate of grafted plants.

Table 1: Comparison of yield between grafted v/s non- grafted plants

Vegetable	Yield Increase (%)	Reference
Melon	3.4 -92	Salar et al, 2015; Mohammadi et al, 2015; Esmaeili et al, 2015
Watermelon	22.7-43.0	Mohamed et al, 2014; Soteriou et al, 2015
Cucumber	8.8-57.0	Farhadi and Malek et al, 2015; Gao et al, 2015
Pepper	9.2	Jang et al, 2012
Aubergine	27.7	Gisbert et al, 2011
Tomato	5.4-80.3	Wahb-Allah 2014; Bhatt et al, 2015; Suchoff et al, 2015
Artichoke	21.7	Temperini et al, 2013

Source: CAB International 2017. Vegetable Grafting: Principles and Practices (G. Colla, F. Perez-Alfocea and D. Schwarz)

4.2 PRESENT STATUS OF GRAFTING

According to a survey conducted by the Japanese government, the utilisation of grafted plants in Japan exceeds 90% for watermelon and cucumber, 79% for eggplant, and 58% for tomato (NARO Research Institute of Vegetable and Tea Science- Japan). It is highly popular in Greece, where the ratio of production area of grafted plants is 90-100 percent for early watermelon cropping, 2-3 percent for tomato and egg plants, and 5-10 percent for cucumbers (Khah et al, 2006). With 129 million grafted seedlings produced, Europe and Spain lead the way, followed by Italy (47 million grafted seedlings) and France (28 million grafted seedlings) (FAO, 2009). According to a survey of 16 Italian nurseries with around 80% transplant output, less than 10 million grafted plants were produced in 1999, which has climbed to more than 60 million plants in 2011. (Leonardi, 2016). China is the world's biggest producer of cucurbitaceous and solanaceous veggies. Grafted plants account for around 40% of watermelon, 20% of melon, 30% of cucumber, 15% of eggplant, 1% of tomato, and 1% of pepper. Grafting is also utilised in the cultivation of bitter gourd and wax gourd. Similarly, in the Netherlands, all tomato plants grown in soilless circumstances are grafted tomato transplants (Bie et al., 2017). Vegetable grafting is becoming more popular over the world, notably in Eastern Europe, North and South America, India, and the Philippines. In China, over 1,500 commercial nurseries are producing grafted transplants. Some countries such as Canada export grafted transplants to Mexico, thus the international trading of grafted vegetable transplants is rapidly increasing (Bie et al., 2017).

Table 2: Main Countries where vegetable grafting is Popular

Continent	Countries
East Asia	China, Japan, Korea, the Philippines
Europe	Spain, Italy, the Netherlands, France, Greece, Cyprus, Belgium, Portugal, Germany, Croatia, Bosnia and Herzegovina
Middle East & North Africa	Turkey, Israel, Morocco, Egypt, Iran, Algeria
America	Mexico, Canada, the USA, Argentina

Indian scenario

In India, vegetable grafting is a new concept and not so popular like fruit crops. Only few institutions have started systematic research work on this innovative technique. IIHR, Bengaluru has initiated research work in tomato for waterlogged conditions; TNAU has started working on cucurbits. CSKHPKV-Palampur has initiated systematic research work on vegetable grafting and has evaluated, screened and identified 30 rootstocks of chilli, brinjal, and tomato

Table 3: Countries position of vegetable grafting in world

Country	Watermelon	Cucumber	Melon	Tomato	Brinjal	Pepper
Israel	70%		5%	15%	5%	-
Japan	93%	72%	30%	48%	65%	5%
Korea	98%	95%	95%	15%	2%	25%
Greece	100%	5-10%	40-50%	2-3%	-	-
Spain	98%	-	3%	4500 ha	-	-
Morocco	-	-	-	75%	-	-
Cyprea	80%	-	-	170 ha	-	-
Italy	30%	-	5-6 million	1200 ha	-	-
France	-	3%	1000 ha	2800 ha	-	-
Netherlands	-	5%		50%	-	-
Turkey	30%	5%		25%	10%	-

Source: Histil South Africa (PTY), 2007 cited by Yassin and Hussen for bacterial wilt, nematode and gummy stem blight resistance. A semiautomatic grafting machine has been installed in Department of Vegetable Science and Floriculture, CSKHPKV-Palampur to accelerate vegetable grafting teaching and research work. Some private companies are also working in vegetable grafting in India.

4.3 VEGETABLE GRAFTING PROCESS

Four main steps of Grafting are as follows:

1) Selection of rootstock and scion variety/ cultivars
2) Seedling production, creation of union by physical manipulation
3) Healing of the graft combination
4) Acclimation of the grafted plants and transplanting of grafted plants in field or protected structure (Lee and Oda 2003; Lee et al., 2010).

FLOW DIAGRAM OF VEGETABLE GRAFTING PROCESS

Selection of scion and rootstock

↓

Seedling production

↓

Grafting

↓

Healing

↓

Acclimatization

↓

Transplanting of grafted plants in field or protected structure

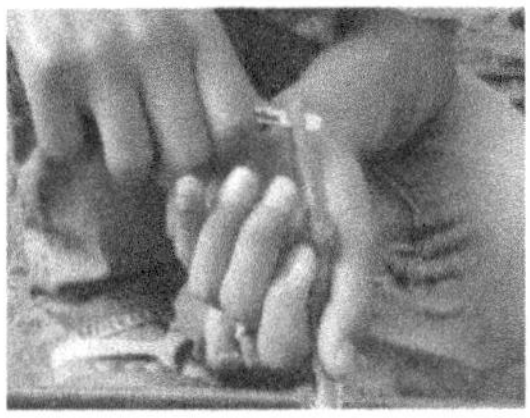

4.4 ROOTSTOCK AND SCION SELECTION CRITERIA

Success rate of any grafted plant depends on selection of appropriate planting material/rootstock. There are different parameters for scion and rootstock which can be taken into consideration at the time of selection process of material for grafting. Seed selection criteria for scion are dependent on genetic purity, genetic viability, yield, and market demand. However, the seeds of rootstock cultivars should be pure, viable, wider adaptability in different climatic zones and soil, with good compatibility and shows resistance for biotic and abiotic stresses.

4.5 BENEFITS

Table 4: Benefits of vegetable grafting

Crop	Benefits	References
Tomato, Pepper, Cucumber, Aubergine, Watermelon and Melon	Enhanced fruit quality	Rouphael et al., (2010), Lee et al., (2010), Zhao et al., (2011), Gisbert et al., (2011), Condurso et al., (2012), Krumbein and Schwarz (2013); Tsaballa et al., (2013); Verzera et al., (2014) and Kyriacou et al., (2016)
Cucumber	Improvisation in scion vigor	Jeong (1986) and Lee et al.,(2010)
Cucumber	Encourages reproductive growth of plant	Jeong (1986) and Lee et al.,(2010)
Tomato	Showed tolerance to flooding	Lee et al., (2010) and Bhatt et al., (2015)
Tomato, Watermelon and melons	Upsurge the nutrient uptake	Lee et al., (2010); Colla et al., 2010b, 2011); Huang et al., (2013b, 2016a,b) and Nawaz et al., (2016)
Tomato	Resistant to Nematode	Dong et al. (2007); Lee et al., (2010) and Louws et al., (2010)
Tomato, cucumber, pepper, watermelon	Improvise the salt tolerance	Colla et al., (2010a, 2012, 2013); Huang et al., (2010); Fan et al., (2011); Yang et al., (2012, 2013); Wahb-Allah (2014); Xing et al., (2015) and Penella et al., (2015)
Tomato, Pepper	Manage drought tolerance	Lee et al. (2010), Schwarz et al. (2010), Wahb-Allah (2014) and Penella et al. (2014)

[Table Contd.

Contd. Table]

Crop	Benefits	References
Tomato, Pepper, Cucumber	Manage high and low temperature tolerance	Venema et al.(2008); Li et al. (2008), Lee et al. (2010); Schwarz et al. (2010) and Lopez-Marin et al. (2013)
Tomato, eggplant, pepper watermelon, cucumber and melon	Disease resistance to soil borne and foliar pathogens	Bletsos et al. (2003, 2005, 2006); Sakata et al. (2008); Lee et al. (2010) and Shibuya et al. (2015)
Melon	Enhancement in post-harvest life	Zhao et al.(2011)
Tomato, Pepper, cucumber and melon	Enhances yield	Lee et al. (2010); Gisbert et al. (2011); Temperini et al. (2013); Moncada et al. (2013) and Tsaballa et al. (2013)
Tomato, cucumber	Tolerance to heavy metals and pollutants	Zhang et al. (2010, 2013) and Kumar et al. (2015a,b)
Cucumber	expansion of harvesting period	Dor et al. (2010) and Louws et al. (2010)
Tomato, Capsicum and Cucumber	Improved Yield and quality Bacterial wilt, nematode and gummy stem blight resistance	Kumar et al. (2018)

CHAPTER - 5

DISEASE MANAGEMENT IN VEGETABLE CROPS THROUGH RESISTANT ROOTSTOCKS

Grafting plays a remarkable role in combating various stresses caused by pathogens and nematodes. Some problems are very serious and their management through chemicals is not effective, moreover resistant varieties are scanty and sometimes not available to the growers. Various fungal pathogens such as Fusarium, Phytophthora, and *Didymella solanacearum* are very destructive in nature. They cause complete crop failures in certain cases. Other serious problem in cucurbits and solanaceous vegetable is root knot nematodes. Root knot nematodes are difficult to manage. Now a day's policy makers and planners are forcing on either organic agriculture or natural farming practices. They advocate no use of chemical on crops, and moreover many chemicals which are used to control/ manage bacterial/ fungal or other pest problems are now banned for use. Under these circumstances only option left with growers to use technologies like Japan and Korea. Vegetable grafting is a best alternate under such conditions. The use of resistant rootstocks is becoming popular all over the world. We have to identify and screen resistant/ tolerant rootstocks.

5.1 FUSARIUM WILT

Fusarium species have a profile capacity to recognize and re-introduced into the infested soils in the production area via equipments, plant based tissues and contaminated soil (Louws et al., 2010). Among vegetable crops several crop species were used as strategic plan for combating the incidence of Fusarium spp., alike in brinjal scions the incidence of Fusarium wilt (*F. oxysporum f. sp. Melongenae*) limited by utilizing the resistant rootstock species of brinjal i.e. *Solanum torvum* (Gousset et al., 2005). But on the basis of success rate to managing the Fusarium wilt incidence through grafting is found to be more pronounced in cucurbits than that of solanaceous crops.

In the matter of subject to minimizing the incidence of diseases non-host resistance through exploitation of host-pathogen technique using inter-generic and inter-specific kind of grafting strategies has been proven as the most considerable management practices, specifically in case of crops belong to cucurbitaceous group. The best example of inter-specific grafting hybrids was mentioned in watermelon, where bottle gourd found to be most effective against the pathogens of watermelon and used commonly in Japan to get disease free crops as alternative approach (Sakata et al., 2007). A rootstock name "Shintoza" found to be effectual against all the races of *fusarium sp.* with promising stability in crop yield in comparison to non-grafted seedlings of the crop (Miguel, 2004). Meanwhile, grafting approach gained popularity in Japan, especially in cucumber resistance for *F. oxysporum f. sp. cucumerinum* was combined with the rootstocks which ensure the tolerance for environmental stress and increase the yield of the crop at same time (Sakata et al., 2008); for cucumber *Cucurbita ficifolia* considered as an excellent source shows resistance against Fusarium wilt with bonus low temperature tolerance, due to which preferred for winter season cultivation (Lee and Oda 2003). Other rootstocks namely "Shintosa" ensures the resistance for wilt and tolerance for heat stress; whereas, *Cucurbita moschata* showed resistance for Fuasrium wilt and Phytophthora blight likewise and *Sicyos angulatus* L. (Burr cucumber) provide resistance from wilt and root knot nematode, but show susceptibility for gummy stem blight and damping off (Takahashi and Kawagoe, 1971; Sakata et al., 2008).

5.2 VERTICILLIUM WILT

Grafting is one of the easiest and alternative techniques used in tomato and brinjal to manage Verticillium wilt compared to less common in production system of cucurbit crops. Among the various species of verticillium a particular species i.e. *V. dahlia* found to be more serious in vegetable crops on a wide range of diversity, because of its biological and diversity basis. The major carriers of pathogen inoculums could be contaminated plant material, equipments and the most frequent one is soil. Pathogen most frequently appears in a wide range of diversity with greater genetic diversity of production.

Thus, the tomato rootstock found to be effective against the Verticillium wilt disease management, as mentioned in numerous reports with positive response, few interspecific hybrids of Solanum were utilized as rootstock this includes [ISHs; *S. lycopersicum x Solanum sp.; Solanum lycopersicum x S. habrochaites; Lycopersici esculantum x L. hirsutum*] and also referred KNVF-K for corky root rot, N for root knot nematodes (RKN), V for Verticillium and F for FOL resistance (Lockwood et al., 1970; Liu et al., 2009).

5.3 PYRENOCHAETA

Corky root rot (*Pyrenochaeta lycopersici*) is one of the most deadly diseases that affect tomato and eggplant plants, especially in colder soils; vegetable grafting has been proven to be beneficial in managing this terrible disease. For example, in eggplant, a scion grafted on the rootstock of ISH's "Brigeor" F1 hybrid reduces the incidence of corky root rot (Ioannou, 2001), while another rootstock, "Beaufort," has been adopted by organic tomato producers in the nation "Sweden" to reduce the occurrence of this disease (Hasna et al., 2009). The another pathogen i.e. *Phomopsis sclerotioides* casual organism of black root rot and wilt, which majorly shows symptoms on fruits of most of the cucurbit crops and also infects the rootstock of bottle gourd used in grafted watermelon as a source of resistant or tolerant rootstock on the threaten basis to the watermelon crop production (Shishido et al., 2006). So, to overcome the devastating impact of this particular disease in cucumber ISHs, *Cucurbita ficifolia* and *Benincasa hispida* among other selections can be used as rootstock plant material.

5.4 PYTHIUM

Pythium is most commonly present devastating pathogen, which on the huge basis impacts the large portion of the agricultural production system. The maganaement practices including cultural and chemical found best option to mitigate the *Pythium* disease; the cultural practices limits the excess moisture level of the soil, helps to avoid the excess accumulation of salts and to minimize the percentage of other injuries and soil temperature. Whereas, the application of fungicides or fumigations can helps to reduce the intermittent effect of numerous *Pythium* diseases. Apart from this, no other reliable method considered on a priority to find out the specific *pythium* resistant rootstock. In an experiment, results confirmed that squash rootstock offers resistance to watermelon from the *P. debarvanum*, which lacks in when combines with bottle gourd rootstock. Resistant rootstocks in brinjal are wild species, namely *S. xanthocarpum, S. gilo, S. indicum, S. khasianum, S. sisymbrifolium, S. testiculatum* and *S. torvum* likewise, cultivars Pusa Bhairav, Florida beauty and Florida Market (Singh et al., 2015).

5.5 BACTERIAL WILT

Bacterial wilt is the most frequently occurring bacterial disease, caused by *Ralstonia solanacearum* in solanaceous vegetables, with major negative effects on growth and yield parameter of the crops in many tropical and sub tropical regions; due to which management focused on alternative technique adaptation such as vegetable grafting in solanaceous crops. With respect to the management of bacterial wilt in tomato a fair range of tolerant or resistant rootstock open pollinated line selection has been done and offered accessions such as, Hawaii-7996, Hawaii-7998 (*S. lycopersicum*) and CRA-66 (*S. lycopersicum var. cerasiformae*) used on worldwide basis (Tikoo et al., 1979; Black et al., 2003; Rivard and Louws, 2008). For e.g. tomato scion grafted on the rootstock H-7996 and CRA-66 had found with no witling symptoms as compared to the non-grafted heirloom line with incidence percentage of 75 to 79; On worldwide, these rootstock have been considered as most promising source of resistance against numerous numbers of strains of bacterial wilt pathogen (Matsuzoe et al., 1993; Jaunet and Wang, 1999; Lin et al., 2008). In an experiment initiated in CSKHPKV, Palampur rootstock of chilli PI-201232 showed resistance against bacterial wilt in capsicum as compared to the brinjal rootstock ; besides this for bell pepper another assured rootstock for resistance are Surajmukhi, Pant C-1 and AVPP0205 as per Kumar et al., 2016. In case of brinjal, Taiwan offered a rootstock confirmed the highest level of resistance for a wide range of strain diversity i.e. EG-203 followed by another rootstock Hawaii-7996 as mentioned in by Lin et al., 2008. The eggplant rootstock was offered by AVRDC as resistance source to combat the problem of bacterial wilt in tomato production if flooding conditions are present in area, if not then Hawaii-7996 is preferred (Black et al., 2003).

5.6 ROOT KNOT NEMATODE

Meloidogyne incognita is a causal organism, from which produce galls on roots. It is most commonly attack all the vegetable groups as it lives in soil. Infected soil, planting material etc. are the major source of epidemics of root knot nematode in any geographical region. To encounter this problem, grafting can be used as alternative tool in integrated pest management practices as grafting technique can reduce the use of fumigants and other chemical hazards at greater extent (Kubota et al., 2008). According to Kaskavalci et al., 2009 tomato scion grafted onto rootstock 'Beaufort' reduced the incidence of root knot nematode significantly at greater extent. Similarly an experiment was initiated by Charles et al., 2012 and reported that hybrid rootstock were used for grafting in tomato, minimize the root galling as compared to non-grafted plants. Whereas, TX301, Aloha and Multifort rootstock can be recommended to control gall formation (Burelle et al., 2016). In addition to this, *Datura stramonium* (Jimson weed) also used as a promising resistant rootstock for tomato, sweet pepper and eggplant (Lowman and Kelly 1946).

5.7 GUMMY STEM BLIGHT

It is caused by *Didymella bryoniae*, known fungal disease and most devastating in cucurbits such as in muskmelon. To cure this serious disease fungicides are used in abundant amount, to control, the use of chemicals is an effective alternative method is vegetable grafting. While damage caused by gummy stem blight can reach upto 100%, leads to severe crop losses, when control measures are not in adequate manner (Vida et al., 2004a, 2004b). Gasparotto et al., 2016 conducted an experiment on four different kind of melon hybrids namely, Sunrise, Louis Bonus II, and Prince hakusho grafted on rootstock of squash hybrid "Shelper" hybrid (immune to *Didymella bryoniae*) and results revealed that significant amount of reduction in gummy stem blight growth was observed in all four hybrids of muskmelon, when grafted on 'Shelper' squash rootstock. In grafted inodorus melon rootstocks RS-841, Elsi and P360 were found highly resistant to *D. bryoniae*, absence of low leaf disease index and crown lesion were noticed especially in greenhouse conditions (Crino, 2007).

5.8 VIRUS PATHOGENS

Vegetable grafting is known for the most common approach to analyze the virus transmission; as more the rootstock susceptibility, more will be the risk of the virus infection in the scion. So, in present scenario, to limits the transmission of the virus to scion introgression of resistance genes into rootstock is foremost aim to be set as a future goal. In addition to this, the etiology of "collapse" mentioned for the tomato has not been elaborated, but found to be in association with the PMV, which can be transferred in nutrient solution from various vectors such as through root grafts and fungal pathogens. The yield losses due to PepMV can be reduced by utilizing the interspecific rootstock through grafting approach compared to non-grafted seedlings as confirmed with findings of Miguel, 2004b; Schwarz et al., 2010.

List of Rootstocks screened for Bacterial wilt and Root knot Nematode at CSKHPKV, Palampur

1. Reaction of Rootstocks for *M. incognita*

S. No.	Crop	Line/Varieties	Reaction	RKI range after 45 days of nematode inoculation
1.	Tomato	Hawaii 7996	-	2 to 4
		Hawaii 7998	-	4 in all
		Palam Pink	#	5 in all
		Palam Pride	#	2 to 5
		CLN 2123-A-1	#	2 to 5
		VI-0343845	+	2 in all
		VI-047335 (EG-195	*	2 to 3
		VI-045376	-	4 to 5
2.	Wild Brinjal	Arka Keshav	-	2 to 4
		Torvum Vigor (*S. torvum*)	*	1 to 3
3.	Chilli	AVPP0205	+	2 in all
		PI-201232	+	2 in all
		VI-037556	+	1 to 2
		Pant C-1	+	2 in all
		97443	+	1 to 2
4.	Capsicum	DPW Rcap-2 No	+	2 to 4
		DPW Rcap-5 No	-	4 to 5
		PB-631	+	1 to 2

(Resistant (+), Mild resistant (*), Susceptible (-), Highly Susceptible (#))

[Table Contd.

Contd. Table]

2. Reaction of Rootstocks for Bacterial Wilt

S. No.	Rootstock	Field conditions	Pots conditions
1.	Tomato		
2.	Hawaii 7996	+	+
3.	Hawaii 7998	+	+
4.	Palam Pink	+	+
5.	Palam Pride	+	+
6.	CLN 2123-A-1	+	+
7.	Brinjal		
8.	Arka Keshav	+	+
9.	*S. torvum*	-	-
10.	VI034845	+	-
11.	EG- 195	+	+
12.	EG-203	+	+
13.	Chilli		
14.	AVPP 0205	-	-
15.	PI-201232	+	+
16.	PB-631	-	-
17.	DPWE Cap	+	-

(Resistant (+), Mild resistant (*), Susceptible (-), Highly Susceptible (#))
Kumar et al., 2015, 2016a, 2016b, 2016c and 2017)

Rootstocks screening under Field conditions

Rootstocks screening under Lab conditions

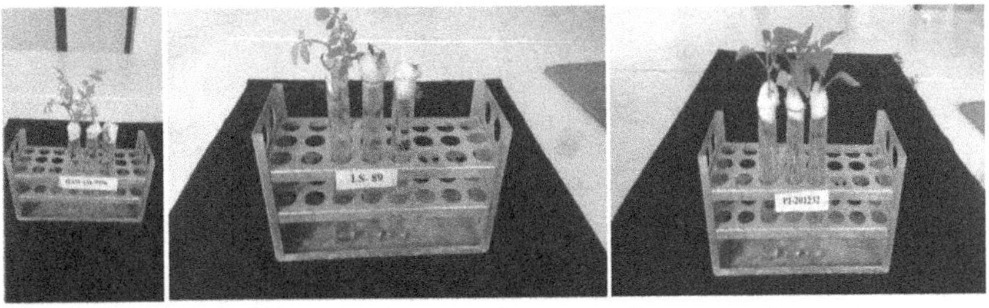

Rootstocks identified for bacterial wilt and nematode resistance

Hawaii 7996 - Bacterial wilt resistance (AVRDC-Taiwan)

Hawaii 7998 - Bacterial wilt resistance (AVRDC-Taiwan)

Palam Pink - Bacterial wilt resistance (CSKHPKV-Palampur)

Palam Pride - Bacterial wilt resistance (CSKHPKV-Palampur)

Arka Nidhi (Brinjal) - Bacterial wilt resistance (IIHR-Bengaluru)

Arka Keshav (Brinjal) - Bacterial wilt resistance (IIHR-Bengaluru)

EG 195- Bacterial wilt and nematode resistant (AVRDC-Taiwan)

EG 203- Bacterial wilt and nematode resistant (AVRDC-Taiwan)

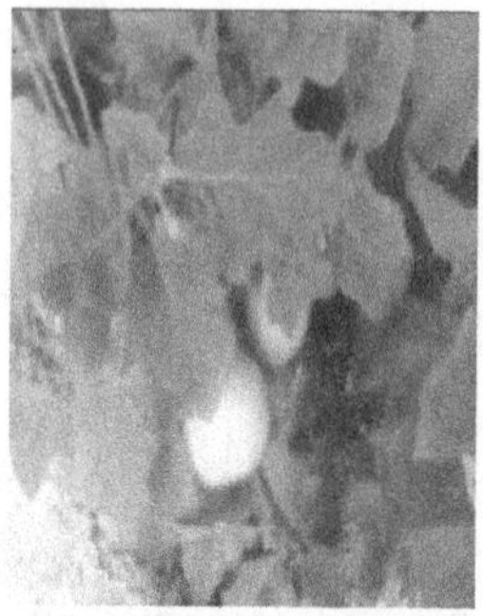

TS000003- Bacterial wilt and nematode resistant (AVRDC-Taiwan)

***Solanum torvum*- Bacterial wilt and nematode tolerance**

5.9 CONCLUSION

Grafting is a rapid tool to the relatively slow breeding methodology aimed at increasing biotic and abiotic stress tolerance of vegetables. It provides a site specific management tool for soil borne diseases. It fits well into the organic and integrated crop production system. It also reduces the need for soil disinfectants and thereby environmental pollution. The use of grafting as an integrated pest management tool to manage biotic stress will be most successful when carried out with increasing knowledge about the biology, diversity and population dynamics of the pathogen and when complemented with sustainable farming system practices.

CHAPTER - 6

INFRASTRUCTURE AND GRAFTING AIDS

6.1 SCREEN HOUSE

A screen house is mostly used to grow of seedlings preceding to grafting and hardening of seedlings preceding to transplanting. To eliminate the transmission of viruses through insects such as whiteflies and aphids screen house should be constructed by using 60-mesh nylon netting material. The upper half of the screen house construction, must be encased with a separate layer of UV resistant or transparent polythene material. About 50% of shade net ought to be sited about thirty centimeter overhead the peak point of the house to lessen the temperature and light intensity. In addition to that double door, the system prevents the entry of insect and pest inside the screen house and sidewise reduces the chance of diseases as well. If insects are discovered within the screen house, prompt management methods should be implemented, taking into account the bug population and the type of the damage. Furthermore, partial shade inside the screen house may be required for the period of the first two or three days, when grafted seedlings are exposed out of the grafting chamber for hardening purpose. Next to that a screened ventilation ridge next to the top of the house is attached for houses of six centimeter or can be additional in width. Ultimately vent lower down the heat that accrues in large screen houses.

Screen House

6.2 HEALING/ GRAFTING CHAMBER

Grafted seedlings are kept in healing chamber to avoid wilting and accelerating healing process. Vegetable grafting requires high relative humidity (85-90%) and ideal temperatures (25-30 °C) during grafting to limit scion transpiration until rootstock and scion vascular tissues mend together and water transport is restored. This occurs in a healing chamber where humidity, light, and temperature may be controlled. Grafted seedlings are maintained in this chamber for an average of 5-8 days to allow the rootstock and scion to develop a solid vascular connection.

Construction Process of Healing Chamber

Select a site with flat elevation and not subject to flooding. Pipes or bamboo stakes of 30cm length and 1.5-1.8 cm inner diameters are pushed 15cm down into the earth to build a therapeutic chamber. Stakes should be placed every 50 cm throughout the length of the chamber. Bend PVC pipes are placed from either end into the bamboo posts on either side to build the arching frame of the chamber.

Following that, the tunnel floor is coated with a 0.15mm thick black polyethylene sheet. The film's edges are twisted up and secured the tunnel structure, allowing it to contain water. Rows of bricks are laid down on the floor to support plant trays above the waterline.

The primary covering is a translucent or UV-resistant polyethylene sheet with a thickness of 0.1mm, which helps to maintain a high level of humidity within the home and prevents rainfall infiltration. During the healing period, this sheet is covered with two layers of shade nets, with the topmost layer being a silver reflecting net. Plastic clips are used to hold the sheet and nets to the skeleton, while the interior door is coated with clear polyethylene. A 2.5 × 4.0 m grafting chamber will contain around 30 trays (40 × 60 cm), each hosting 40 plants in 6.0 cm diameter pots. As a result, the tunnel chamber has a capacity of around 1,200 seedlings.

High Cost Healing/Grafting Chamber

Low Cost Healing/Grafting Chamber

On the basis of requirement healing chamber can be constructed in a two ways. A simple, low cost grafting chamber can be constructed at farmer's field whereas for commercial purposes a hi-tech chamber of size 50 m^2 or more can be constructed to facilitate the basic requirement of plants such as temperature and humidity.

6.3 ACCLIMATIZATION CHAMBER

After the establishment of a grafted union in grafting chamber hardening of the grafted seedlings is another step that is considered to increase the success rate of grafting, and it can be performed just before the transplanting to eliminate the chances of wilting and leaf burn in grafted seedlings. Acclimatization chamber is used to practice hardening operation in grafted seedling and it takes about 7-10 days for acclimatization as hardening treatment.

Acclimatization is essential for plant life. It is performed to promote healing of the cut surface and hardening of grafted seedlings in order to enhance survival chances when grafted plants are put in the field or greenhouse. It is critical to maintain optimum moisture content before and after grafting in order to produce consistent grafted seedlings. Acclimatization can be accomplished easily by wrapping the rootstock and scion in a black plastic bag (to prevent heat buildup) until the union forms. Acclimatization is typically accomplished by the use of plastic covering by growers. Many commercial nurseries place grafted plants, generally in cell trays, on a greenhouse bench and seal the trays with a single sheet of semi-transparent high-density polythene film 0.01mm or thinner to reduce moisture loss and keep the trays sealed for 5 to 7 days without excess watering. Partial shade, on the other hand, may be necessary throughout the day to minimize excessive heat buildup.

Acclimatization Chamber

6.4 GRAFTING AIDS

Vegetables are tender and succulent; therefore special care is needed for grafting especially in vegetable seedlings. Numerous grafting aids are available in the market to unite 'scion' and 'rootstock' together, such as grafting clips, tubes, pins, blades, tapes, etc. Farmers still prefer to use a single razor blade as a grafting knife. Among all of these grafting aids, many commercial growers would most likely utilise elastic tube-shaped clips with or without attachment to provide support to the grafted seedlings; this approach is widely accepted for both human and robot grafting. The clips, despite their somewhat different sizes, may be reused several times. Furthermore, in the case of tomato and pepper grafting, several nations, notably Israel and the Netherlands, employ a large number of tiny elastic slit-tubes.

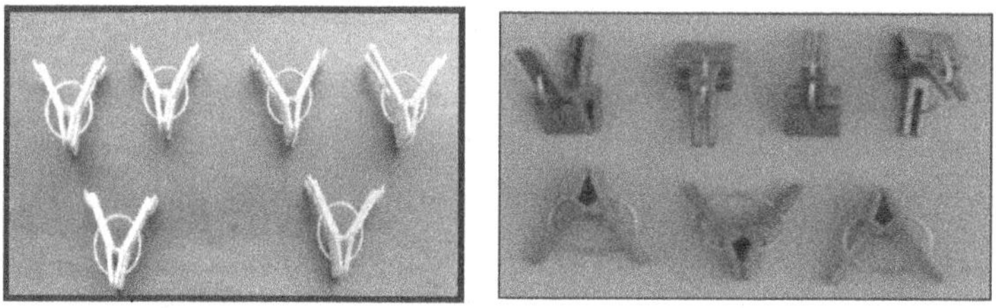

Grafting Clips

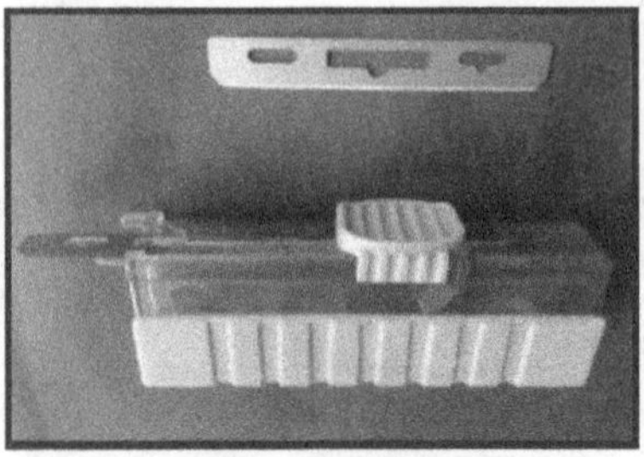

Grafting Blades

Silicon Tubes (1.5-2.0mm)

The ceramic pin is a very useful and effective tool for securing the grafted interface, and it is ideal for machine or robot grafting. It may be used with just hands, a basic pencil-shaped device, or machines and robots. Another method for retaining the grafted counterparts in place is to use adhesive tape or glue, or even aluminium foil. Grafting was formerly done by growers, but because to the efficiency, simplicity of post-graft care, and recent development of the commercial seedling sector, it is now fast migrating to cooperative enterprises (such as sales of plug-grown seedlings).

6.5 ROBOTIC GRAFTING

Grafting is an extremely laborious & time-consuming procedure, especially when it is performed manually, so various measures have been made to mechanize grafting since 1987. To speed up the process of grafting in a short period and to minimize the labour expanse, robots have been engineered which can speed up the process by ten times, without compromising the efficacy, accuracy, and quality of the conventional method.

To improvise the sources of resistant rootstock and for quality improvement, identification of new markers & utilization of the

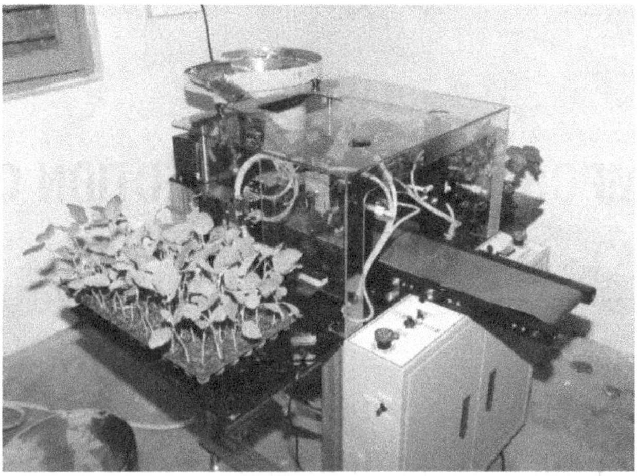

Semiautomatic grafting machine installed at CSKHPKV-Palampur

existing markers both towards combating the biotic and abiotic stresses will be done in improving our rootstock sources & efficiency in grafting compatibility. Many propagators are interested in automated grafting or grafting robots since labour management is a key concern in high scale propagation (more than 10 million plants produced per year) or seasonal propagation. The first semiautomatic cucumber grafting system was commercialized in 1993, with a single semi-automated grafting robot generally grafting at a speed of 600-800 grafts per hour (speed equivalent of 5, 6 skilled workers for cucurbit, & 2-3 skilled workers for tomato), but inspecting the quality of grafted seedlings requires a minimum of two workers & one technically skilled worker.

CHAPTER - 7

IMPORTANCE AND SELECTION OF ROOTSTOCKS

7.1 IMPORTANCE OF ROOTSTOCKS

The precise practices opted by growers for vegetable production such as practiced the monoculture and year around cultivation system, may favors the rapid and high density build-up of soil borne pathogen inoculums, which makes soil sick and results in epidemics of disease in a large geographic area. To prevent the spread of pathogen, particularly under the protected conditions; pre-soil fumigation with microbes and occasionally with other soil fumigants like formaldehyde can be applied as per requirement or threshold frequency of disease incidence. However, in recent period of time continuously upsurging level of chemical use or dependence by farmer's affecting the ecological balance throughout the worldwide. So, to encounter the harmful effects of chemical usage and to develop ecofriendly sustainable soil- borne disease management strategies for protected environment, use of resistant rootstocks is emerging as one of the finest options to deal with environmental stresses.

Resistant rootstock could be utilized as best alternative in order to combat the biotic and abiotic stresses, as resistant rootstock enhance the plant vigour and increases the uptake of nutrients and improvise the water uptake, which upsurge the tolerance of resistance capacity of the plant to minimize the incidence of soil borne pathogen, temperature stresses, salinity and waterlogged conditions etc. The kind of rootstock also affects the performance of the scion such as growth, fruit quality and yield. Due to appropriate exploitation of the vigorous root system of the rootstocks, the grafted plants generally express increase in water and nutrient uptake as compared to non-grafted or self-grafted plants. In addition to it, grafting in plants boosted up the uptake of micronutrients such as nitrogen and phosphorus. In case of tomato rootstock of *Solanum melongena* shown positive response and provide better protection against bacterial wilt (*R. solanacearum*) as well as for flood conditions, particularly in hot wet seasons of South- east Asia (Black et al., 2003). On other hand another wild species closely related to cultivated species of brinjal *Solanum torvum* is one of the species which is extensively utilized as rootstock for tomato and aubergine respectively (Bletsos et al., 2003; Lee et al., 2010; Moncada et al., 2013). Vegetable grafting found to be an efficient approach to resolve the problem associated with disease development in various vegetable crops, such as bacterial wilt and phytophthora blight and the disease resistance is ensured by involving the resistant rootstocks. Jang et al. 2012 studied that a pepper scion Nokkwang grafted on breeding lines PR-920, PR-921 and PR-922 showed maximum survival rate and resistance to bacterial wilt and phytophthora blight. Moreover on that, in Japan bottle gourd is the most important rootstock against Fusarium wilt in watermelon (Sakata et al., 2007). Grafting approach found to be most popular in Japan since the resistance for *F. oxysporum* f. sp. *cucurmerinum* was united with environmental stress tolerance and yield enhancing rootstock (Sakata et al., 2008). For cucumber the rootstock of fig leaf gourd showed an admirable resistance for fusarium wilt with benefits of tolerance to low temperature stress, thus mostly recommended for winter season cultivation as confirmed with findings of Lee and Oda, 2003. In addition to the tolerant rootstock sequence of the cucumber crop another rootstock "Shintosa" ensures the resistance against incidence of fusarium wilt in combination with the heat tolerance mechanism, due to which preferred for the summer season cultivation of cucumber crop. Another rootstock *Cucurbita moschata*, offered tolerance to diseases including phytophthora blight and fusarium wilt; and *Sicyos angulatus* L. (burr cucumber) ensures resistance for root knot nematode and fusarium wilt, meanwhile found to be susceptible for gummy stem blight and damping off where cucumber used as scion material (Sakata et al., 2008).

7.2 ROLE OF ROOTSTOCK IN VEGETABLE GRAFTING

Success rate of grafted plant depends on rootstock; a good rootstock should always shows compatibility with scion and provides desirable results without reducing fruit quality and yield. Rootstock enhances the plant vigour, selected rootstocks should be stress resistant and location specific, vigour root system increases the nutrient and water uptake efficiency and rootstocks induced transformation in scion.

7.3 USE OF RESISTANT ROOTSTOCKS IS THE KEY TO BIOTIC AND ABIOTIC STRESSES

Plants exposed develop a variety of physiological and pathological problems, resulting in stunted development and significant losses in fruit quality and output. Grafting vegetables on rootstocks capable of decreasing the influence of external pressures on the shoot is one technique to avoid or decrease production losses caused by bad soil chemical and physical conditions and environmental stresses. In our country there are many biotic as well abiotic stresses which cannot be managed or controlled through chemicals or which are very difficult to control, only alternative under such circumstances is the use of grafting technology which is being followed in many countries to counter biotic and abiotic stresses.

7.4 ADVANTAGES OF ROOTSTOCKS

The rootstock has several major functions:

1. It anchors the plant in the ground & supports the up-right growth.
2. Upsurge the uptake of water & nutrients for plant growth & fruit production.
3. Root system of the plant stores energy as carbohydrates.
4. It provides a level of tolerance to plants against soil borne diseases.
5. Rootstock allows the scion to adapt in a variety of soil conditions.

Beside all these responsibilities, each rootstock has the characteristics to varying degree such as vigour, compatibility with scion, disease tolerance, nematode tolerance, drought tolerance, cold hardiness, tolerance to salinity and its effects on scion, fruit size, and quality etc.

7.5 SELECTION OF ROOTSTOCKS

The suitable rootstock and scion cultivars must be chosen to ensure the success of grafted vegetable production. At the two true leaf stage, choose the ideal rootstock and scion. The scion's stem diameter should be the same as the rootstock's.

A variety of high-yielding scion and rootstock cultivars with desired traits, including as resistance to diseases and nematodes and tolerance to salt, drought, flood, heat, and cold stress, have been introduced by the public and commercial seed corporations. Seed firms understand how crucial it is that the rootstocks they develop do not degrade fruit flavour or quality. As a result, gardeners have a diverse choice of cultivars from which to pick rootstocks for Cucurbitaceae and Solanaceae crops based on their specific needs. Cucurbit rootstock breeding is mostly centred on China, Korea, and Japan. Because of the growing popularity of grafted plant production, the number of registered cucurbit rootstocks is constantly expanding (Kato and Lou 1989; Ko, 1999; Lee et al., 2008). According to one estimate, around 600 Cucurbitaceae rootstocks are in various stages of development in China, with just a handful issued each year (King et al., 2010).

Recommended Rootstocks for Solanaceous and Cucurbit crops

Rootstock	Purpose (Tolerance)
1. Tomato	
S. lycopersicum	Alkalinity, Thermal shock, virus and improve vigour
S. lycopersicum x S. lycopersicum	Improve bacterial wilt and nematode resistance, fruit quality, macronutrient uptake
S. habrochaites	Low temperature, resistant to corkey root disease and yellow leaf curl virus

[Table Contd.

Contd. Table]

Rootstock	Purpose (Tolerance)
S. lycopersicum x S. habrochaites	Salinity, abiotic stresses, improve fruit quality, vigour and yield, macronutrient uptake, management of soil borne disease, root- knot nematode, *Colletotrichum coccodes* and *Sclerotium rolfsii*
S. lycopersicum x *S. cheesmaniae*	Enhance quality and performance in salinity conditions
S. lycopersicum x	Potassium and phosphorus deficits,
S. pimpinellifoilum	drought, salinity and soil impedance, Improve water use efficiency
S. lycopersicum var. *cerasiforme*	Management of soil borne and foliar pathogens, arthropods and weeds
S. melongena	Alkalinity, bacterial wilt, low and high temperature, Enhance flood stress resilience
S. lacinatum	Resistant to water logging
S. chilense	Tomato yellow leaf curl virus
S. pennellii	Tomato yellow leaf curl virus
S. pimpinellifoilum	Tomato yellow leaf curl virus
S. integrifolium	Enhance fruit quality
S. luteum	Alkalinity
Nicotiana tabacum	Alkalinity
Datura patula	Alkalinity
S. sisymbrifolium	Enhance fruit quality and disease resistance
S. toxicarium	Enhance disease resistance
S. nigrum	Enhance fruit quality
Rootstock	**Purpose (Tolerance)**
2. Brinjal	
S. melongena x S. melongena	Early, resistant to nematode, graft compatible
S. torvum	Reduce cadmium concentration, resistance to Verticillium wilt, nematode and salinity

[Table Contd.

Contd. Table]

Rootstock	Purpose (Tolerance)
S. incanum	Nematode resistant, graft compatible
S. incanum x *S. melongena*	Nematode resistant, graft compatible
S. melongena x *S. aethiopicum*	Nematode resistant, graft compatible
S. macrocarpon	Increase fruit quality and yield
S integrifolium	Resistant to fusarium wilt and verticillium wilt, reduce cadmium concentration
S. sisymbrifolium	Resistant to verticillium wilt
S. torvum x *S. sanitwongsei*	Resistant to bacterial wilt, high temperature
S. lycopersicum	Increase efficacy to verticillium wilt resistance
S. lycopersicum x *S. lycopersicum*	Nematode resistant, graft compatible
S. lycopersicum x *S. habrochaites*	Nematode, salinity resistant, graft compatible

CHAPTER - 8

POTENTIAL VEGETABLE CROPS FOR GRAFTING

8.1 INTRODUCTION

Vegetable grafting is utilized mostly in solanaceous and cucurbitaceous crops. Among all the solanaceous crops like tomato, capsicum and brinjal are the most important crops in which grafting is practiced worldwide due to its commercial value. But in cucurbits, cucumber is the only cucurbit grown on commercial scale in India apart from this grafting technique is also utilized in crops such as watermelon, bitter gourd and muskmelon to impart resistance from various biotic and abiotic stresses. Due to high market demand of cucurbitaceous vegetable, they are cultivated intensively in the same land both under open field and polyhouse conditions.

8.2 GRAFTING OF SOLANACEOUS AND CUCURBITACEOUS CROPS

1. Tomato

Grafting in tomato has been started in about 1960 to 1970. Grafting technique is becoming more popular among the tomato growers, not just because it helps to minimize the incidence of root disease but also improve the fruit quality and yield through enhanced plant vigour and stronger root system. Tomato grafts showed their best performance to abate the incidence of fusarium and bacterial wilt. Nevertheless, it not only upsurge the yield and vigour of plant but also provide resistance to number of pathogens such as fusarium and verticillium sp. In the context of mid hill of HP bacterial wilt in tomato is most of the overwhelming disease in open and protected conditions. Number of rootstocks has been identified to prevent or to decrease the incidence of bacterial wilt especially for mid hill regions of HP.

Grafted Tomato Seedlings

2. Bell Pepper

Bell Pepper is most important crop mostly grown as an off-season vegetable under protected environment. But production of bell pepper under polyhouse conditions in Himachal Pradesh is very low as compared to national and international levels. The reason for low production under polyhouse conditions in Himachal is lack of suitable bacterial wilt resistant varieties or hybrids. Apart from this other biotic and abiotic stresses also affects the bell pepper production. In Himachal Pradesh bacterial wilt was first time has been seen in 1981, Kangra valley on solanaceous

Grafted Capsicum

vegetable crops and in 1985 reached at sporadic level, and now present at endemic level not only in Kangra but in Mandi district as well. Researchers attempts number of chances to develop a successful resistant hybrids but no resistant hybrid is available till date. In sweet pepper, grafting is considered to be an effective approach to combine the resistance with desirable horticultural aspects, the best example of this is, grafting of sweet pepper scion on chilli and sweet pepper rootstocks offers the management opportunity by minimizing the chances of soil borne disease (Phytophthora Blight, bacterial wilt) and flooding related problems.

3. Eggplant

Eggplant (*Solanum melongena* L.) is a major vegetable crop consumed at international level and has number of nutritional and medicinal properties. Among all members of the solanaceae family, the first use of grating technique was recorded in the 1950s with eggplant grafted on scarlet eggplant (*Solanum integrifolium* Poir.). The eggplants, *Solanum integrifolium* ensures the higher resistance percentage for bacterial wilt than for fusarium wilt among all other resistance source species/ cultivars of eggplant and remains vigorous until the later stages of eggplant production; making it the most common among the available rootstock in Japan. In Asia and Europe, eggplant can be grafted on

Grafted Eggplant

eggplant and on torvum (*Solanum torvum*), eggplant (*S. melongena*), and scarlet eggplant (*S. aethiopicum*) and a variety of rootstocks are commercially available. In a commercial scale production of eggplant the yield losses at huge level often recorded with the incidence of the soil borne diseases like verticillium wilt, bacterial wilt, root knot nematode and fusarium wilt and to control the drastic effect of these diseases a rootstock "*S. torvum*" showed resistance promising towards all mentioned diseases with enhanced vigour ability of the scion plant portion.

4. Cucumber

Grafted Cucumber Seedling

Cucumber (*Cucumis sativus* L.) is grown for its edible tender fruits in almost all parts of the world. In India, cucumber solely cultivated in an area of 18,000 hectares with a production of 1,20,000 tones. Grafting can be successfully practiced in cucumber grafted onto kinds of various cucurbits species, such as *Cucumis spp.*, fig leaf gourd (*C. ficifolia* Bouche), *Cucurbita spp.*, *Cucurbita* interspecific hybrids, and *Luffa* sp. and bottle gourd. The most important reason for using grafted cucumber plants is to ensure the resistance from fusarium wilt and enhances the plant ability of low temperature stress tolerance. Fig leaf gourd is one of the best rootstock material can be preferred for cucumber that ensures the greater ability to resist the fusarium wilt attack, make sure the enhancement in cold tolerance mechanism of the plant. Apart from this, interspecific hybrids of *Cucurbita* had gained the popularity for summer cucumber grown in summer season, to provide heat tolerance along with long-lasting Fusarium resistance. Fig leaf gourd (*Cucurbita ficifolia)* is unique amongst all cucurbits species having low temperature tolerance of up to 15^0 C showing 6^0 C lower than that of cucumber roots.

Cucumber grafted onto fig leaf gourd, an excellent rootstock grows faster than cucumber with own rooted system and produce summer squash even at low temperature conditions, the root system of this rootstock shows significant ability to enhance the water and nutrients absorption efficiency. In H.P. parthenocarpic cucumber varieties are being cultivated under polyhouse conditions. Downy mildew is the serious disease being noticed. Though, parthenocarpic varieties do not have any serious problem in the state except nematode infestation. Low temperature rootstocks may be useful in providing tolerance to the varieties grown during winter season.

CHAPTER - 9

GRAFTING TECHNIQUES IN VEGETABLES

9.1 INTRODUCTION

Vegetables are tender and succulent in nature, though various grafting techniques are being used to graft the vegetable seedlings. Method of grafting in crops depends on the nature and type of the crop and also on availability of facilities. Generally, grafting techniques can be grouped into two categories i.e. manual and mechanical or robotic grafting. In manual grafting almost all the grafting process is executed manually and while in mechanical grafting, machines (robots) are used to perform the main grafting processes. Meanwhile number of grafting machines or robots are present to facilitate convenient services to a farmer, by minimizing the cost of labour and time saving procedure, but still in many countries manual grafting is found to be most convenient method to develop grafted unions of vegetable crop plant.

9.2 GRAFTING METHODS OF VEGETABLES

1. Tongue Approach/Approach Grafting
2. Cleft Grafting
3. Hole Insertion/Top Insertion Grafting
4. One Cotyledon/Slant/Splice Grafting

5. Tube Grafting
6. Pin Grafting
7. Micro-Grafting
8. Automated Grafting

1. Tongue Approach/Approach Grafting

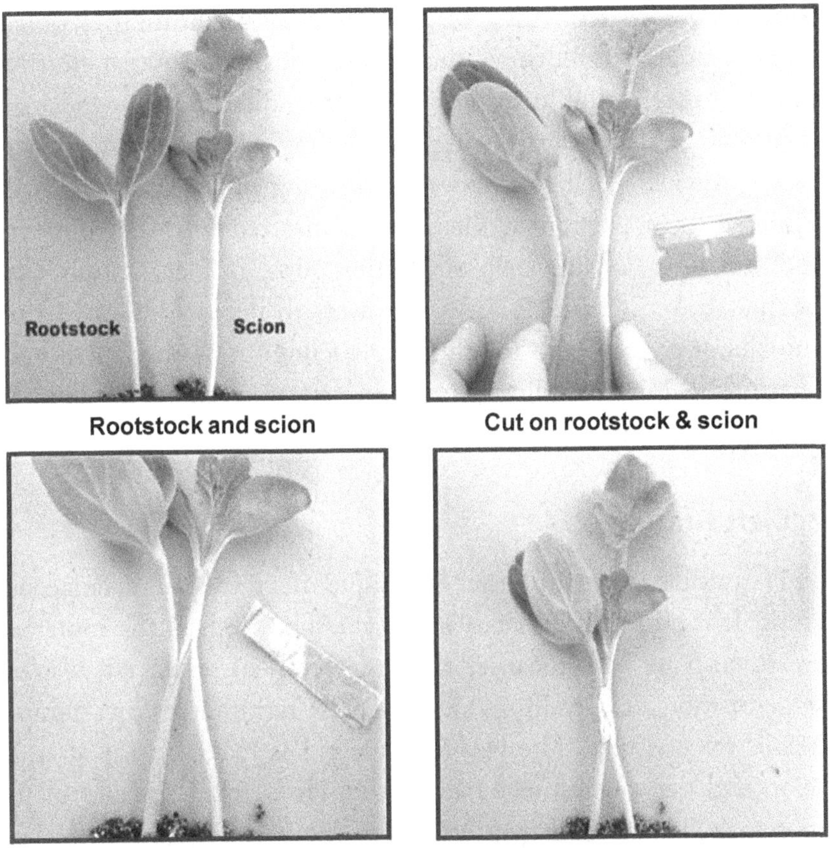

Rootstock and scion **Cut on rootstock & scion**

Joining of scion to the rootstock Fixing the joint with a grafting strip

Various stages of Tongue Approach Grafting

(**Source:** Jinger, D., Dhar, S., Gupta S. and Parashar, A., 2014)

This grafting is mostly used for cucurbitaceous (cucumber, melon) and sometimes used in greenhouse grown tomato when healing & acclimatization are less than favorable for successful tube grafting. For successful grafting scion and rootstock should be of same height and diameter. Scion seed needed to be sown as four to five days earlier than rootstock seed material. The one cotyledon and growing point of the rootstock should be detached before grafting to avoid the overcrowding of the plants in the healing chamber. The cut on grafted rootstock should be made on downward direction and upward direction in scion to the perpendicular axis at the angle of 30^0-40^0. After that growing point along with one cotyledonary leaf is also detached to avoid the crowding of the excess growth. Then, the cut portion of the scion is inserted into the rootstock and attached with the help of grafting clips. It is commonly used by the farmers and small nursery growers. It needs additional space and labour as compared to other approaches but has higher survival rates of seedling. A skilled person can graft approximately about 80 grafted unions in a day (Lee et al., 2010).

2. Cleft Grafting

This grafting is broadly used technique of grafting in solanaceous crops. It is effortless and easy approach. The seeds of the rootstock are sown 5 to 7 days earlier than the seeds of scion. At 4-5 leaf stage of rootstock seedlings are cut at the right angles by retaining 2-3 leaves on stem. The wedge shape cut is made on the stem of scion and the tapered end fixed into a cleft cut in the end of the rootstock. The graft is grip hard with the help of plastic clip. It is suitable for rootstocks with wide hypocotyls.

Cleft Grafting

3. Hole Insertion /Top Insertion Grafting

Top or hole insertion grafting is widely used in cucurbits such as watermelon. This approach is preferred mostly when; rootstock and scion have hollow kind of hypocotyls. In hole insertion method sowing of rootstock seeds are done 3-4 or 7-8 day prior to the watermelon seeds, particularly in case of squash and bottle gourd rootstocks. Whereas in case of solanaceous crops rootstock nursery is sown 5 to 10 days before scion and grafting is done after interval of 20-25 days. In grafting process firstly, growing point along with leaves are detached from rootstock followed by slant hole is made through a wooden or plastic gimlet and then scion is prepared having slant cut with tapering end, insert the scion into rootstock.

This method is more convenient to perform than other methods due to establishment of strong and compatible union and vascular connection. In this technique a person in a day can graft more than 1,500 grafts. To accomplish the higher percentage of achievement rate, needed to maintain the temperature at $21-36^{0}C$ and relative humidity approximately at 95% till transplanting of the seedling material in the field conditions.

Root stock and Scion

Hole preparation on the scion

Fixing the joint with the clip

Clipping the joint

Various stages of Hole Insertion Grafting

(**Source:** Jinger, D., Dhar, S., Gupta S. and Parashar, A., 2017)

4. One Cotyledon/Slant/Splice Grafting

The commercial seedling nurseries have recently been adopted this technique. One cotyledon grafting technique is relevant to most vegetables. This grafting has been devised for robotic grafting or can be performed manually as well. In this method, one cotyledon leaf and growing point of rootstock are removed followed by slant cut is made at the angle of 35^0 to 45^0 and then scion is attached with the help of grafting clips, pins or tubes. After grafting, the plants should be kept in the dark at 25^{0}C and have to maintain 100% relative humidity for three days.

Rootstock and scion material

Cut on the rootstock and scion

Fixing the root stock & scion

Clipping the joint

Various stages of Slant Grafting (Source: Maurya, A.K., 2014)

5. Tube Grafting

Tube grafting is analogous to slant grafting with the exception of that in this technique root stock and scion joints are detained with an elastic tube as a substitute of clips. The seeds for rootstock are sown 1-2 days prior to seed for the scion. Since smaller plants are used in tube grafting, we must be able to graft plants 2-3 times more rapidly than with the more usual cleft technique. The smaller plants occupy less space during the healing & acclimatization process. In this technique the tube may be used several times depending upon the materials. It is commonly used in tomato and brinjal.

Tube Grafting (Source: Maurya, A.K., 2014)

6. Pin Grafting

Pin and splice grafting technique have all the same steps that has to be followed for successful grafting, but for holding the rootstock and scion graft instead of using clips like in splice grafting specifically intended pins are used to hold the grafted plant union. This pin is made up of ceramic about of 15 mm long and diagonal width of hexagonal cross section of 0.5 mm and left in plant without and harmful effect. Ceramic pins are more expensive than clips and it can be used only once at a time, on other hand it save time and labour because pins are not required to remove once it is attached to the grafted union. To make this method cheaper and more effective instead of using ceramic pins, rectangular shaped sticks of bamboo/wooden could be used as alternative of ceramic pins.

7. Micro-Grafting

Micro-grafting refers to in-vitro grafting using micro-explants or very small parts from meristematic tissues. This grafting technique is usually expensive and conceded out to eradicate the viruses from infected plants, as virus particles do not present in the apical meristem.

8. Automated / Mechanical Grafting

- In this technique of grafting a simple machine can generate about of 600-700 grafts/hour with two operators, while manual grafting practices produce approximately of 1,000 grafts/person/day (GU, 2006).
- A completely automated grafting robot generates 1000 grafted plants/hour with a 90-93% success rate.
- At present total six models of fully automated and semi-automated grafting machines (robots) are available in different international markets, out of which three models have been introduced by Japan and one each by Spain, Korea and the Netherlands, respectively.

9.3 ADVANTAGES AND DISADVANTAGES OF DIFFERENT GRAFTING METHODS

Method of grafting	Advantages	Disadvantages
Hole insertion grafting	Strong vascular connection	Trained workers required
	Grafting clips not required	Grafting machines are not available
	Less labour intensive	removal of offshoots is essential

[Table Contd.

Contd. Table]

Method of grafting	Advantages	Disadvantages
	During the healing period not necessary to plant scion with rootstock	Controlled environment is required for healing process
	Less space required for healing	Only used in cucurbits
Tongue grafting	Not required trained labour	Weak vascular connection
	Grafting machine available	Grafting clips required
	Controlled environment is not needed	More labour intensive
	Performed in solanaceous and cucurbits	Offshoots removal is required
		Scion & rootstock are required to be planted together during the healing process
		Extra space essential for healing
Splice, Tube or One- cotyledon grafting	Very strong vascular connection	Trained workers required
	Grafting machine available	Grafting material like clips, tubes and pins required to fix the graft union
	Off-shoots removal is not required	More space required for healing
	Scion does not required to be planted with the rootstock during the healing process	More labour intensive
	Performed in solanaceous, cucurbits and other minor crops	Controlled environment is required for healing process

(Colla et al., 2017)

CHAPTER - 10

GRAFTING TECHNIQUES IN TOMATO

10.1 INTRODUCTION

Tomato is one of the commercially grown crops and number of soil borne diseases, drought and salinity stress incidence have been observed in this crop, which distresses its quality and yield. To encounter these problems grafting is emerging as an alternative method, tomato scion is grafted on different rootstock of crops, so the vigorous root system can improvise the plant metabolic activity to encounter the different kind of stresses.

Brinjal as a rootstock

- Utilized to manage the soil borne diseases such as fusarium wilt, bacterial wilt and root knot nematode (Yamkawa, 1982; Goth et al., 1991; Kalloo, 1993) and in those regions where chance of water logging or flooding is higher, in addition to this, the combination of eggplant lines and tomato lines could be utilized in more efficient manner to obtain a successful grafted plant.
- Selection of the eggplant rootstock will be done on the basis of two foremost criteria i.e. quality and maximum return in terms of yield of the scion.

- Promising rootstock are needed to utilized in a more prominent manner, e.g. *S. torvum* shows significant resistance and provide a good vigor for scion, to meet out the soil borne diseases namely bacterial wilt, fusarium wilt and root knot nematode (Yamkawa, 1982; Ginoux & Laterrot, 1991).

Tomato as a rootstock

- Tomato rootstock can be utilized in those areas where the problem of flooding and water logging is not found.
- Selection of the rootstocks can be done on the basis of the resisting the impact of bacterial wilt & other soil borne diseases, e.g. The rootstocks recommended by AVRDC tomato lines Hawaii 7996 and Hawaii 7998 for the reason that they possess high resistance against Bacterial and Fusarium wilt.

HAWAII 7996

HAWAII 7998

- There are two most prominent and popular cultivars of tomato utilized on commercial scale particularly in U.S. namely Beaufort and Maxifort (released by DeRuiter Seeds, Bergschenhoek,

The Netherlands) showed resistance for nematode, Verticillium Wilt, Fusarium root rot, Crown rot, Corkey rot, and tomato mosaic virus. The major difference among these two cultivars is that 'Maxifort' delivers more strength to the scion as compared to 'Beaufort'.

10.2 GRAFTING METHODS IN TOMATO

- Tube grafting
- Cleft grafting
- Tongue grafting

Tube Grafting

The following strides exemplify the procedure of grafting of tomato scions on eggplant rootstocks. Almost similar procedures will be followed to graft scion of tomato on tomato rootstock.

1. Select the eggplant rootstock and tomato scion and having same diameter of stem (1.6-1.8 mm). To accomplish this, sowing of eggplant nursery will be performed about three days before sowing of tomato.

2. Make a cut on the rootstock at the angle of 30^0 above the cotyledon. Then commence the cut as high as possible on the stem.

3. Cut the scion stem at a 30^0 angle, a little above the cotyledons or first true leaf of stem. Choose a position on the stem to make a cut on a scion to attain appropriate breadth.

4. Slide the 10 mm long latex tube about 2.0 mm from inner diameter and made a cut at the angle of 30^0 over the scion. Ensure that the cut angles of the tube and scion are parallel to each other. Insert the scion approximately half way into the tube.

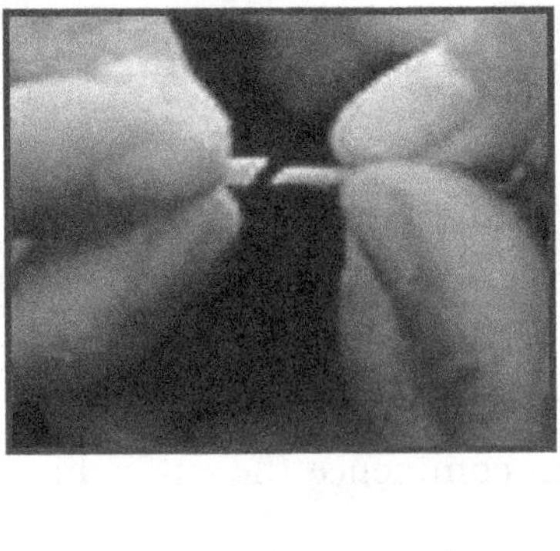

5. Slide the scion above the rootstock seedling. Over again ensure that the cut angles of the tube & rootstock are parallel to each other.

6. Now insert the scion and rootstock jointly so that both touch each other. The tube or a clip will hang on the seedling till the natural hardness of splits and become ready to transplant in the field.

(**Source:** Chen, W.Y., Lin, L.J., Luther, G.C. 2016, World Vegetable Centre, Grafting Manual: How to produce grafted vegetables. Shanhua, Tainan, Taiwan, pp. 1-6)

7. Shift the grafted seedlings without delay into the shaded compartment. Maintain the 25-32^{0}C temperature in the chamber, with a thin layer of water on the polyethylene floor liner and relative humidity of 85 % by keeping the doors closed. Place the trays of seedlings on bricks to hold up the plants on top of the water line. Initially the grafted seedlings start wilting but will become normal and upright within three days.

Low cost grafting chamber

8. Start the hardening process by peeling away the top (silver) layer of shade net material after 4-5 days of grafting. Empty the floor pan by draining water. Open the plastic covered door of chamber however to prevent insect infestation keep the screen door closed. Maintain such environmental conditions for 2-3 days.
9. Shift the seedlings to screen house from the chamber. After nine days of grafting, spray the grafted seedlings with 0.5% urea solution, or 2 g/litre of N:P: K (19:19:19) or the equal of a similar water soluble fertilizers. Then hold the grafted plants in the screen house atleast for 7-8 days for advance development and hardening. This complete process takes about 30-33 days for successful establishment of a grafted plant from sowing.

Cleft Grafting

1. Seeds for rootstock material are sown 5-7 days prior to the scion seed.
2. The stem of the scion and rootstock attaining 2-3 leaves are cut at right angle.
3. The scion stem is cut in a wedge, and the tapered end of scion fitted into a cleft cut in the end of the rootstock.
4. With the help of plastic clip the graft union is held firmly.
5. Shift the grafted plants to healing chamber to ensue healing up.

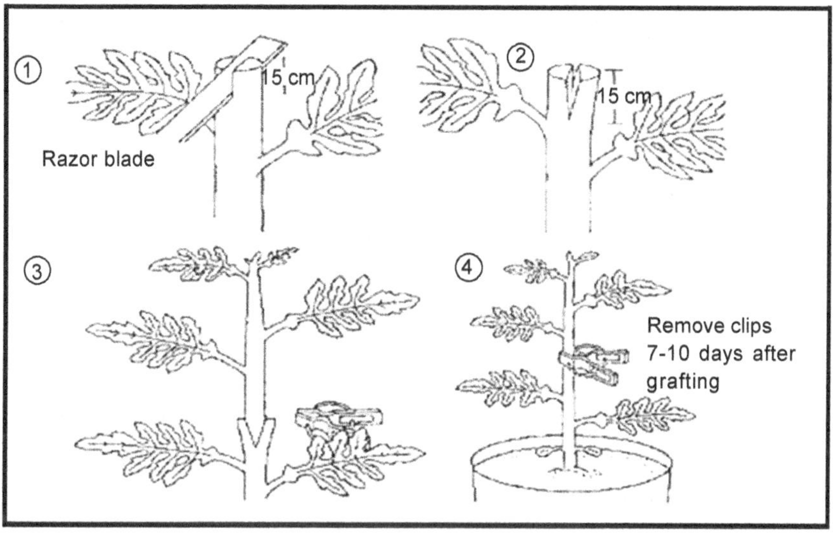

Various Stages of Cleft Grafting (Source: Mohanta, S., Prasad, BVG., Rahaman, S. and Bareily, P., 2015, Vegetable Grafting)

Tongue Grafting

1. Perform a downward cut at a sharp angle of 60 degree relative to the horizontal plane, a half way across the rootstock hypocotyl and upward cut on same angle in scion hypocotyl.
2. Clasp the upward 'tongue' of rootstock hypocotyl with the downward 'tongue' of scion hypocotyl.

3. The graft union holded with a grafting clip or by covering with a short strip of lead tape.
4. Plant them collectively in a container or a large cell.
5. After seven days in healing chamber, cut the shoots of rootstock and roots of scion.
6. Remove the clip or tape from seedling.

10.3 CASE STUDIES IN GRAFTED TOMATO

- An experiment was conducted at CSKHPKV, Palampur on effect of root stocks and grafting techniques on yield and quality of tomato (*Solanum lycopersicum* L.) under protected environment. The results revealed that, Brinjal rootstock VI-047335 grafted onto tomato scion (GS-600) showed most prominent results for most of the horticultural and quality parameters. Moreover all the rootstocks found with least incidence of bacterial wilt as compared to control and these rootstock are Palam Pink , Palam Pride, VI-045376, VI-047335, VI-034845, Arka Nidhi and Arka Keshav. Whereas, slant cut followed by cleft grafting was recorded as most prominent approach for grafting particularly in tomato, on other hand the cost benefit ratio was found much higher with grafted plants under protected conditions than non-grafted ones (Amirullah et al., 2017).
- In an investigation conducted at CSK Himachal Pradesh Krishi Vishvavidalaya Palampur, HP, grafting has been performed in tomato to analyze the impact of various rootstocks on different traits and the results confirmed that the "VI-034845" a brinjal rootstock showed highly significant effects for plant height, no. of fruits, fruit yield and for quality traits viz., pericarp thickness, total soluble solids and ascorbic acid, respectively (Kumar et al., 2017).

- Another experiment conducted by Mokhles et al., 2019, found slant cut grafting method as highly acceptable in tomato along with the brinjal rootstocks VI-047335 and VI-045367 showed improvised quality aspects. Whereas, all the rootstocks confirmed the resistance to bacterial wilt viz., Arka Nidhi, Palam pride, Palam pink, VI-045367, VI-047335 and VI-034845, compared to non grafted plant.

Performance of grafted tomato plants

CHAPTER - 11

GRAFTING TECHNIQUES IN CAPSICUM

11.1 INTRODUCTION

Capsicum is the most important and widely grown off-season vegetable crop under polyhouse conditions in Himachal Pradesh. The major purpose of grafting in capsicum is to diminish the incidence of soil borne diseases, particularly to manage the *Phytophthora capsici*, nematodes and viral diseases. Besides this, scion quality and compatibility is also important so as to combat the major environmental stresses (Gisbert et al., 2010; Chavez-Mendoza et al., 2013; Jang et al., 2013).

11.2 GRAFTING METHODS IN CAPSICUM

- Tube grafting
- Cleft grafting
- Tongue grafting

Tube Grafting

1. Tube grafting in capsicum utilizes Indra (capsicum) as a scion and Surajmukhi (chilli) as a rootstock. Selected scion and rootstock stem with the similar diameter, 1.6-1.8 mm. To attain this, sow the rootstock about 5 - 6 days before scion.

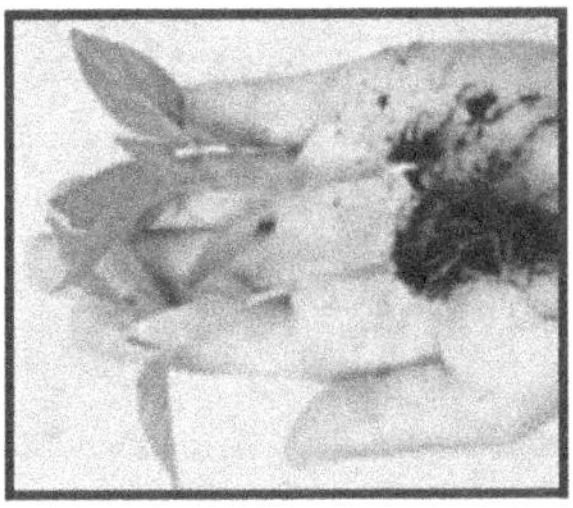

2. After this cut the stem of rootstock at 30° angle and 1.5 cm above the cotyledon, or first true leaf of stem.

3. Then cut scion stem at 30° angle, a little above the cotyledon. It is important that the scion and rootstock stem should have similar diameter. Carefully select a point on the stem to cut the scion to attain the suitable diameter.

4. Over the scion stem slide a 10-mm long latex tube (2-mm inner diameter, cut at a 30° angle). Ensure that the cut angles of the tube and scion are parallel. Now push the scion about middle into the tube.

5. Slide gently the scion (now fitted with the latex tube) over the seedling stem of rootstock. Again, ensure that the cut angles of the tube and rootstock stem are corresponding or parallel.

6. Make sure that the rootstock and scion are in complete contact with each other. The tube should be on the seedling until it naturally hardens the splits and ready to transplant in the field.

7. Seedlings of grafted pepper in tray is now ready for shifting to grafting chamber.

(**Source:** Chen, W.Y., CLin, L.J. and Luther, G.C. 2016)

8. Now shift the grafted seedlings instantly into the shaded chamber maintained at optimum temperatures of 25- 32^{0}C. A thin layer of water is maintained in the floor liner of polyethylene and high humidity (>85% RH) by closing the doors. Keep seedling trays on bricks to sustain the plants above the water line. Initially the grafted seedlings start wilting but become normal and upright within three days.
9. After 4-5 days of grafting, start the hardening process by peeling away the top (silver) layer of shade net material. Drain the water out of the floor pan. Plastic-covered door of the chamber should be open while keeping the screen door closed to avoid insect invasion. Such condition should be maintained for 2-3 days.
10. Shift the grafted plants from the chamber to a screenhouse. Foliar spray of 0.5% urea solution, or 2 g /liter of NPK (19:19:19) or the equivalent of a similar soluble fertilizer should be applied after nine days of grafting. Keep the grafted plants in the screenhouse for 7-8 days for further growth and hardening. After sowing the entire process complets in 50 to 55 days.

Cleft Grafting

1. Cleft Grafting in capsicum is the one in which indra (capsicum) taken as a scion and VI-034845 (brinjal) taken as a rootstock. Sow the seeds of the rootstock 5-7 days earlier than those of the scion.
2. Both stems i.e. shoot (scion) and the rootstock are cut at right angle, by retaining 2-3 leaves on the stem.

3. The stem of the scion is cut in a wedge, and the tapered end fitted into a cleft cut in the end of the rootstock.
4. The joint of the scion and rootstock union is tied with the help of grafting clip.
5. The size of the rootstock should be more than that of the scion.

Tongue Grafting

1. In tongue grafting, Indra (capsicum) taken as a scion and Pant C-1 (chilli) taken as a rootstock and made a downward cut at a sharp angle (60 degree relative to the horizontal plane) half way across the rootstock hypocotyl by using a grafting blade.
2. Make an upward cut at the equal angle a mid way crossway the scion hypocotyl.
3. Cuts are comparable in length, so they can match up together.
4. Both slits are fitted together with a grafting clip or sealed with an aluminum wrap to allow healing to take place.
5. After a week of healing process, cut the rootstock shoot and scion roots off near the graft union.

11.3 CASE STUDIES IN GRAFTED CAPSICUM

- An investigation was conducted at CSKHPKV, Palampur on bell pepper to analyses the effect of rootstock and grafting techniques on various horticultural parameters and incidence of bacterial wilt under protected conditions. PI-201232 has been found most appropriate bacterial wilt resistant rootstock for bell pepper and recorded with positive results for most of the horticultural and quality parameters. Tongue grafting was reported as best grafting method and also expressed high grafting success rate in combinations (Rana, 2015).

- The standardization of robotic grafting has been done in capsicum, where the scion (Indra) grafted on rootstock (PI-201232) of 45 days old enhanced the growth, yield and quality traits of bell pepper. It can be utilized in future to ensure the higher success rate with quality product and ease in management (Sarswat et al., 2020).

Performance of grafted Capsicum Plants

CHAPTER - 12

GRAFTING TECHNIQUES IN CUCURBITS

12.1 INTRODUCTION

Grafting is one of the most convenient method which can be utilized more prominently in cucurbit crops including cucumber and melons particularly watermelon. Moreover, there are number of rootstock which shows compatibility with all three crops, while other rootstocks are more specific to a specific crop. Grafting technique in cucurbits started in Japan and Korea in late 1920's, where scion of watermelon was grafted on squash rootstock (Davis et al., 2008).

More emphasis on grafting in cucurbits is given, just to provide the resistance against nematodes and to minimize the incidence of the soil borne diseases, particularly after the ban on fumigant 'methyl bromide' and pathogen appears with more virulent races, on other hand there is no effect of common pesticides on the virulent races of disease pathogens (King et al., 2008; Davis et al., 2008; Lee et al., 2010). Furthermore, grafting have an effect on vegetative growth, flowering, fruit ripening date and quality and give high production, specially under low- temperature conditions.

12.2 WHY GRAFT CUCURBITS?

- To upsurge the quality and yield of the crop
- To encourage plant vigour
- To develop tolerance to various abiotic stresses
- To develop resistance to soil borne plant pathogens

12.3 METHODS OF GRAFTING IN CUCURBITS

Generally, three grafting techniques commonly used for cucurbits like cucumber, watermelon and other melons are:

- Hole-insertion grafting method
- One-cotyledon grafting method
- Approach grafting method

Hole-insertion grafting method

Hole-insertion grafting approach doesn't need grafting clips to attach the scion on rootstock. Even without it, it can establish the strong grafted union. The major benefit of this approach is that, there is no need of scion and rootstock of same size, scion could be taken as flexible as is required or smaller than of rootstock size. Fix the scion by inserting into the rootstock hole on its top. The orientation of scion cut surface should be monitored. Rootstock tissue must attach with the cut surface of scion and then make orientation of scion in such a manner, so the cut surface faces the downward position. After the process of grafting, grafted union are shifted to healing chamber, because grafted union requires extra care for the strong establishment of the grafting union among two crops or same crop. Healing chamber need to maintain optimum temperature, humidity and light.

Various steps of hole-insertion grafting are:

1. Before grafting, rootstock cuttings are harvested 30-60 minutes early. Cracking can be avoided by using rootstock having less turgid tissue.
2. With the help of scalpel detach the first true leaf, the apical and lateral (axillary) meristems from the rootstock. Near the base of the petiole the axillary meristems appear as white dots otherwise it will result in rootstock shoot growth.

Splice Grafting

3. Puncture at an angle so that the edge of the tool comes out through hypocotyls below the contrary cotyledon. Leave in the perforating tool so that we do not need to 'look' for the hole.

4. Insert the scion shoot into the rootstock hole on its top. Orientation or position of scion cut surface should be monitored. The cut surface must touch the rootstock tissue and adjust the scion that its cut surface faces down.

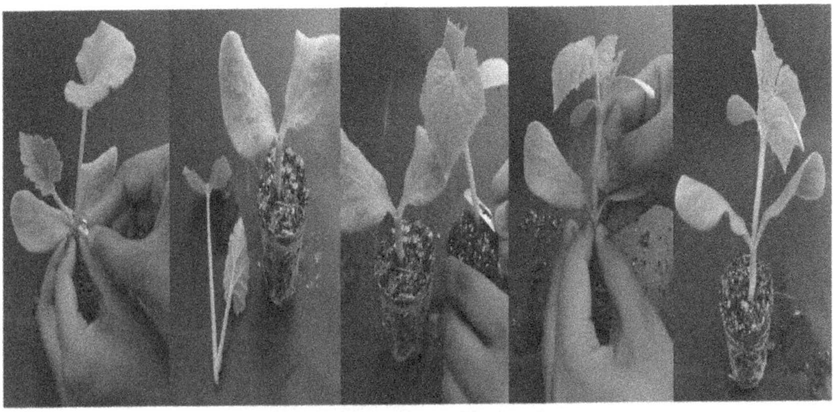

Hole Insertion Method

5. Transfer the grafted cuttings into a fresh tray and then sprinkle water on cuttings to stay away from too much dryness.
6. Thereafter place the grafted seedlings in the tray, shift the seedlings to the healing chamber for up to a week for healing and rooting. After this seedling is ready for transplantation.

One-cotyledon grafting method

This method of grafting is one of the most suitable method developed by Japanese engineers, especially for automation or robotic grafting. With the ease in performing and simplicity of this method, it has become a manual grafting approach. One of the best advantages of this approach is that, it can be done with or without removal of roots form the rootstock. Various steps of one-cotyledon grafting methods are:

1. A cotyledonary leaf should becut from rootstock seedling at 45° angle, jointly with apical and axillary buds.
2. Now get ready the scion with similar hypocotyl thickness, cut at the equal angle approximately at 10-15 mm underneath the cotyledons.
3. Two cotyledonary leaves (one rootstock and one scion leaf) should be united together by grafting.
4. Unite both the plants (scion and rootstock) jointly with the help of plastic clip.

5. Transfer the grafted seedlings to fresh tray. Apply water immediately to avoid dryness.
6. Shift the grafted seedlings for the healing process, for a week.
7. Remove the grafting clip after the union has taken.

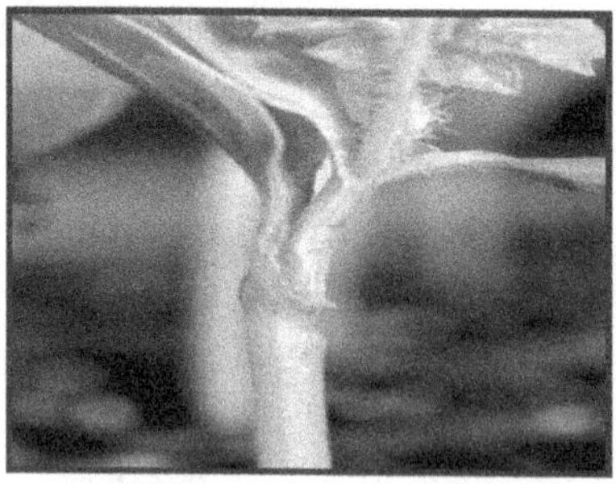

Approach grafting method

Approach grafting method assures the maximum percentage of achievement rate, whereas the speed of grafting of this technique is less fast than rest (two) techniques. Removal of scion roots can be done after the healing is completed, so as a result highest success rate could be achieved even if the healing conditions are not favourable for the grafted plants. Another positive point of this grafting method is that, there have been some chances of flexibility regarding the size of scion and rootstock. In this method there are least chances of shoot development within rootstock particularly in field conditions. So to perform the approach grafting simple steps are as following:

1. Create a downward cut at a sharp angle (60° degree relative to the horizontal plane) mid way crossways the rootstock hypocotyl.
2. Create an upward cut at the equal angle a mid way transversely the scion hypocotyl.
3. Hook the upward 'tongue' of rootstock hypocotyl with the downward 'tongue' of scion hypocotyl. Grip them jointly with a grafting clip or by wrapping with a short band of lead tape.
4. Grow them jointly in big cell/container.
5. After a week in healing chamber cut the rootstock shoot and scion roots off.
6. Remove the clip or tape.

Other grafting methods

Other grafting techniques used for cucurbitaceous plants are pin-grafting and cleft (V-shaped) grafting.

Research Outcome of experiment done at CSKHPKV-Palampur

With respect to achievements enlisted in the account of CSKHPKV, Palampur, in cucurbits particularly in cucumber, successful outcomes have been mentioned by Anjali et al., 2020a. The rootstock of bottle gourd along with the slice grafting method showed significantly highest success rate with enhanced growth

Grafted cucumber plants on various rootstocks

and yield contributing traits. While, in the same year, refreshers reported that the splice grafting with combination of bottle gourd rootstock followed by summer squash and pumpkin rootstocks confirmed the improvisation in horticultural, yield and quality traits and all the rootstocks i.e. summer squash, pumpkin, bottle gourd, bitter gourd, ash gourd, cucumber and Luffa showed resistance against the gummy stem blight (Anjali et al., 2020b).

CHAPTER - 13

POST GRAFT HANDLING MANAGEMENT IN VEGETABLE CROPS

13.1 INTRODUCTION

Success of the grafted plants mostly depends on two essential procedure which have to be followed after the completion of the grafting i.e. healing and hardening. Appropriate healing and hardening are very important for the survival of grafted seedlings. Grafted plants must be sheltered from dehydration and wilting till the grafted union heal jointly. To attain high success rate of grafting, seedlings needs high relative humidity and optimum temperature conditions to decrease transpiration of the scion until the vascular tissues of rootstock and scion are healed jointly and the restoration of water transport takes place. This process occurs in a healing chamber where humidity, light and temperature can be regulated. After five to eight days of grafting, vascular connection between scion grafted onto rootstock takes place and two weeks for healing of graft union properly. Initially the scion is not capable to take delivery of water from the rootstock. It is therefore, important to maintain proper environmental conditions to prevent water loss from the scion and promote rapid formation of the graft union in first week after grafting.

Immediately after the grafting, grafted seedlings are shifted to healing chamber in order to prevent the grafted seedlings from water stress. This can be accomplished by lowering the transpiration stream-the movement of water from inside the plant tissues into the atmosphere and to maintain this high humidity, low temperature and light is required. But beside this, while lowering down the temperature, growth at the joining tissues possibly will slowed down. Therefore, the main focus of effective chamber process is elevated humidity and no exposure to direct sunlight at the same time as the grafted union are in developing stage within two to four days after the grafting. Meanwhile during grafting scions are seperated out from their roots, due to which scion portion starts showing the wilting symptoms, mostly in the first few hours after the grafting.

After the separation of scion from root system the scion starts wilting in a first few hours after grafting, due to damaged water vascular system, vascular tissue in the rootstock and scion must align in such a way so that their tissues can easily develops together by forming a strong union for water and nutrient uptake.

13.2 POST GRAFT HANDLING MANAGEMENT IN VEGETABLES

1. **Construction of healing chamber-** Healing chamber can be constructed inside the greenhouse, sheds or either in shady areas. The frame of the healing chamber is made up of blue PVC pipes of one inch, which works well for the frame construction especially of healing chamber with combination of Drape clear UV stabilized 200 micron polyethylene covering sheets over the frame and to cover the edges on the three sides of plastic with soil so as to prevent the moisture loss from the chamber. Provide entrance door on the fourth side of the

healing chamber. To block out the sunlight cover two layers of shade cloth in combination with one layer of black polyethylene sheet.

2. **Temperature, humidity and sunlight in healing chamber-** For the proper healing of the grafted seedlings/ plants inside the chamber, it is required to maintain minimum exposure to direct sunlight, 80- 95% of humidity and a temperature range from 25 to 30^0 C.

Healing conditions: Temperature-25-30 ^{0}C, RH-80-95%, Low light intensity

3. **Healing chamber management-** To maintain the water availability for the rootstock and scion, water them atleast 12-24 hours before grafting. Frequently mist the plant while performing grafting through hand sprayer. Whereas, in addition to that a few hours earlier to grafting, mist the healing chamber indoor area with water either by using spay nozzle, hose or spray bottle (preferred to a very small volume chamber), spray the water inside the chamber until the four walls, ground and ceilings are get wet and water start accumulating on the bottom side of the chamber.

To lower down the temperature if the temperature has risen up outside, misting can also be done. However, once grafted seedlings /plants are kept inside the heal chamber, close the chamber compactly to avoid the moisture loss. But the excess of moisture content inside the healing chamber may lead to upsurge the water on the leaves of the scion which could also leads to de-attachment of the scion from the rootstock. To get rid out of this situation, do not mist the grafted plant once it is placed inside the chamber, as to avoid the damage from the overhead irrigation, apply water from the bottom during the healing process.

4. **Healing the grafted plants:** To avoid the transplanting shock while introduced to the greenhouse environment, adjust the exposure time for grafted plants. Nevertheless, management of the grafted chamber environment is the procedure which has to be taken into consideration and the following points are:
   i. Do not disturb the healing chamber atleast for two days and seal it properly.
   ii. On 3rd day after grafting open the healing chamber just with enough opening space so water can be sprayed easily inside the plastic, in order to upsurge the humidity then close it again.
   iii. Avoid the water accumulation near by the graft unions as it may lead to incidence of disease, necrotic tissue and failure of grafts. If these symptoms appear on grafted plant, stop misting plants directly.
   iv. On 4th day do not disturb the chamber.
   v. On 5th day, open the healing chamber just for 30 minutes, to spray the water on the plastic surface and then close it.
   vi. Then, repeat the same procedure on 6th and 7th day, open chamber for one hour and for six to eight hours respectively.

vii. On 8th day remove the grafted plants from the chamber followed by replenish the carbon dioxide through opening the healing chamber.

5. **Acclimatization**

- To acclimatize the grafted plant, maintain the healing chamber environment without light and high humidity atleast for two to four days, during this period scion return to the normal turgor level with no evidence of moisture stress.
- Once the grafted plant reaches to the normal turgor level, slowly introduce it to light and consequently humidity level get reduced over the following weeks.

Healed seedlings kept for acclimatization

- Avoid the direct sunlight exposure when introduced to light for first time in healing chamber and if the chamber is established indoor, fluorescent light could be used above the chamber.
- In greenhouse shading material such as shading nets are used to avoid the direct exposure of sunlight instead of restricting it altogether.

- It is necessary to maintain low relative humidity in case of greenhouse or other location where this chamber is constructed.
- Steadily bring in the grafted plant to the greenhouse environment as to elude the chances of kind of biotic and abiotic stress and to minimize the mortality rate of grafted plants.
- After moving the grafted seedlings to the greenhouse environment and to prevent the any kind of physical damage to the grafted plants these are watered from the bottom side, because graft union are still sensitive.

6. **Problems associated with healing chamber management**

- If the humidity is below the optimum level inside the chamber or grafted seedlings are taken out of the healing chamber too early or immediately, scion leaves start wilting. Since the healing chamber have congenial temperature and relative humidity also have good environment for pathogens and other microorganisms.
- The physiological disorders may be observed if the humidity in the healing chamber is comparatively large or saplings are kept in chamber for longer duration. The graft union of scion develops adventitious roots which are excised with the help of sharp razor blade.
- Sometime other problems may arise due to high humidity like callus growth (edema) on the leaf veins. This problem is not reversible but could be prohibited on new plant if transfered to less humid chamber.
- Shift the grafted seedlings again in the healing chamber and mist them, if wilting is observed.

CHAPTER - 14

TRANSPLANTING OF GRAFTED SEEDLINGS AND AFTER CARE

14.1 INTRODUCTION

The management practices of the grafted and non-grafting plants are almost same in case of all vegetables, but some extra care should be given to grafted plants, till the proper establishment of the union between two plants. Proper healing and acclimatization are very important for grafted plants to survive. After the grafts are made, the plants must be protected from wilting until the cut ends heal together. Successful grafting of vegetables require high relative humidity and temperature following grafting to reduce transpiration of the scion until rootstock and scion vascular tissue are healed together and water transport is restored. This process occurs in a healing chamber where humidity, light and temperature can be regulated. It takes an average of 5-8 days after grafting for the rootstock and scion to establish vascular connection and 14 days for the graft union to heal properly. During the first week after grafting, the scion is unable to receive water from the rootstock. It is therefore, important to maintain proper environmental conditions to prevent water loss from the scion and promote rapid formation of the graft union. The grafted plants are shifted to a healing chamber immediately after grafting. The purpose of the healing chamber is to prevent the scion from becoming water stressed during the process of grafting. This can be accomplished by slowing the transpiration stream-the movement of water from inside the plant tissue into the atmosphere.

To get a successful graft union:

a) **Hardening:** Hardening can be done by exposing the plants to full sunlight and some water stress in the days before grafting. It increases the tolerance of plant to water stress.

b) **Turgidity:** It can be maintained in the plant by providing irrigation a few hours before grafting.

c) **Shade:** A shady location should be selected for taking up grafting operation.

14.2 TRANSPLANTING OF GRAFTED PLANTS AND AFTER CARE

To understand the transplanting of grafted seedlings and rest of the aftercare procedure the following points are needed to take into consideration:

Graft transplanting in the main field

↓

Raised beds and shelters

↓

Depth of transplanting

↓

Removal of suckers and adventitious roots

↓

Staking and pruning of the grafted plants

↓

Water management

↓

Pest management

1. **Graft transplanting in the main field** - Grafted plants are transplanted into the field when they are 6-10 inches in height, at this stage they do not have any flower buds, flowers or fruits. For solanaceous crops, silicon clips or tubes are used in grafting which will fall off as the stem increases in diameter, so removal of these clips will not be needed. Spring plastic clips used for cucurbits may not drop off and they will restrict plant growth if not removed. For both solanaceous and cucurbit crops, remaining grafting clips are to be removed after 2 weeks of transplanting. After appropriate sanitation grafting clips can be reused (e.g., washing in 10% hypochlorite solution and rinsing in water).

Silicon clip drop off after healing of graft union

2. **Raised beds and shelters** - One of the major advantages of vegetable grafting is that grafted plants are recommended particularly for hot-wet season. However, the planting of grafted plants is done in hot-wet season, to avoid hazardous effects of flooding, raised beds are preferred for transplanting along with the combination of rain shelter, particularly made up of clear polyethylene covering material. Rain shelter could be used as

protective shield to the grafted plants, as it provide protection from the abiotic stresses like heavy rainfall, winds etc and also provide shade to lower down the temperature during hot weather conditions. Rain shelters in combination with grafted plants increases the yield particularly in summer months.

3. **Depth of transplanting** - The major precaution taken during the transplanting of the grafted plants is to make sure that grafted union remains above the soil level. The closer the graft union is to the soil line, the more likely adventitious roots from the scion will develop and grow into the soil. In this situation, grafted plant becomes more prone to the incidence of soil borne and other diseases, as pathogen penetrates into the resistant rootstock, which leads to infection and death of the entire plant.
4. **Sucker and adventitious roots removal** - In case of eggplant rootstock removal of sucker that develops near to the cotyledon is needed. While, to prevent the grafted pants from the incidence of soil borne diseases, removal of the adventitious roots develops from the scion, before it contact with soil is done.

Suckering

Planting depth

5. **Staking and pruning of the grafted plants** - To support the grafted plants staking is important in case of crops such as tomato, capsicum and cucurbits and it will be performed after two to three weeks of transplanting. Pruning is one of the best method which allows plants to grow into a more healthier plant by removing of the stem as in case of indeterminate type of tomatoes. In case of tomato and cucurbits only single stem is allowed to grow whereas, in case of capsicum two stem are allowed to grow. Indeterminate tomatoes are pruned as to allow the single or two stems to develop properly. To stake the plant, tie it with thread or any kind of staking material. This will prevent the plant scion or vine to fall down and fruit contact with the soil.
6. **Water management** - Water requirement is one of the factor which affects the growth and development of the plant by affecting the metabolic activities inside the plant, which occurs on the plant as visible symptoms. So, the water management is much needed as per the crop requirement, particularly when grafting is performed, because the sap flow during the grafting is very high, due to which seedling shows wilting symptoms. Misting of the water on the grafted plants is needed. When rootstock is brinjal the water requirement is higher similarly, when scion of tomato is grafted on rootstock of eggplant and the water management is not proper, chances of blossom end rot increases and it can be cured by maintaining the high soil moisture.
7. **Pest management** - Insect pest and diseases are one of the major factors which cause the maximum losses and ruin the crop. Vegetable grafting is used as an alternative technique to cure the incidence of these damaging factors. During the hot-wet seasons, diseases which are more prone in case of vegetable

crops are such as early blight, southern blight, black leaf mould, gray leaf spot, bacterial spot, and tomato leaf curl virus in tomato. Whereas, the major insect which are observed most commonly are tomato fruit worm, tobacco cutworm, beet armyworm, and leaf miner. To avoid the incidence of insect pest and disease monitor your crops closely and take appropriate control measures as per the kind of crop and pathogen.

CHAPTER - 15

PROBLEMS AND PRECAUTIONS ALLIED TO VEGETABLE GRAFTING

Despite green revolution in the country to make India self sufficient in food production, technologies like grafting in vegetables did not become popular among growers of the country. Sometimes it is a routine question by the stakeholders that if grafting has been proved beneficial in Japan and Korea, then why it has not been in practice in a country like India? Many agricultural scientists did not think or consider it as a tool to counter stresses, rather they focused only on developing hybrids or varieties resistant or tolerant. Japan and Korea took lead and they are successful in managing various problems in vegetables. Organic agriculture and safe crop production demand by the consumers is becoming talk of the town and there is a class of people who love to buy organic produce with either no or zero input of chemicals. It is an agreed concept that vegetable grafting is labour intensive and requires a lot of patience. It has numerous advantages as already discussed in previous chapters. Our aim is to promote this technology at a grass root level so that a large number of farmers adopt it and reduce dependence on chemicals to manage various production issues. Our research and extension work is based on practicality of the technology and various myths related to it should be resolved. It is true that in our country very less research work on this aspect has been done in the past, but now even some farmers are also trying to make it profitable and sustainable. There are few problems associated with grafted plants, which can be addressed easily by taking precautions. We have noticed a major issue in transportation of seedlings to a distant place, if grafts are not properly placed in plug trays. Grower's need not to worry about its transportation if seedlings properly healed.

15.1 PROBLEMS ASSOCIATED TO VEGETABLE GRAFTING

1. Grafting technique is labour intensive and requires specialized trained workers.
2. Time management is necessary for sowing of rootstock and scion seeds.
3. It is necessary to provide a controlled environment for the healing process.
4. Lack in the availability of the automatic machines for grafting.
5. Chances of rootstock and scion incompatibly at initial stage and after transplanting.
6. Rootstock and scion incompatibility induces undergrowth or overgrowth of the scion which lead to decreased water and nutrient flow through the grafted union, significantly affects the yield and quality of the scion fruits.
7. Development of offshoots/ suckers, leads to higher incidence of diseases.
8. Grafting can upsurge the risk of pathogens occurrence like seed borne pathogens in the nursery and viral diseases also takes place.

9. The cost of grafted seedling is higher compare to normal vegetable seedlings which is a drawback to adopt grafted seedlings.

Though the problems associated in grafting are few but if not addressed timely then growers may face more problems. Therefore, modern techniques such as use of grafting robots for commercial production may be advantageous for entrepreneurs. The cost of production is one of the major aspect in vegetable grafting as farmer has to buy the seed of rootstock as well as seed of scion. For this purpose farmers are advised or given training to use local available rootstocks duly screened and evaluated by the scientists.

15.2 PRECAUTIONS NEEDED TO BE TAKEN IN VEGETABLE GRAFTING

1. To capitalize on the ability of the technique, an ideal synchronization of the vegetable cycles has to be achieved prior to the conjunction of the two plants.
2. Prior to grafting expose seedlings to full sunlight and a bit of water stress to retain the plants short and augmented tolerance to water stress.
3. To obtain higher rate of success grafting in case of cleft, pin, splice and hole insertion grafting approaches, there is a need to maintain RH about 95% in first 48 hours at temperature varies from 27 to 28^{0}C inside the healing chamber (Guan and Zhao, 2014).
4. Early in the morning or just after dark are best times to graft as transpiration has typically slowed to reduced levels.
5. To avoid chances of contamination use of pest free high quality seeds, sterile trays, knifes and blades.

6. Grafting location protected from direct sun light and apart from the greenhouse heater discharge.
7. To obtain perfect graft plant assures that the scion and rootstock stem diameter and cut angles should be same.
8. Make sure the cut sides of both rootstock and scion unites properly when clipped together as to ensure the chances of compatibility with each other.
9. Utilization of physical and chemical barriers to prevent the incidence of virus vectors, insects and fungi.
10. Controlled environmental conditions like composition of the substrate, sunlight, ventilation, temperature and humidity are necessary to maintain during the entire process of grafting till the established grafted plant is ready to transplant.

CHAPTER - 16

VEGETATIVE PROPAGATION TECHNIQUES IN POTENTIAL VEGETABLES UNDER PROTECTED CONDITIONS

16.1 INTRODUCTION

Vegetative propagation is quite common in a few vegetable crops like garlic, ginger, potato, turmeric *etc.* where the most common method of vegetative propagation i.e. cutting of tuber/stem has been practiced. Capsicum, tomato and cucumber are three principle money spinner crops grown on commercial scale under protected conditions. Finding seeds of the desired hybrids or varieties is a major issue with these crops. Private sector hybrids for protected cultivation are expensive and not available all the times, whereas there is shortage of public sector hybrids and are not available to the polyhouse growers. In recent years, approximately 30,000 polyhouses have been constructed in HP under various state sponsored schemes. Quality planting material is another issue which is faced by the farmers due to increasing area every year under protected cultivation. Though some hybrids or varieties are available which perform better under protected environments, thereby limiting the choice of growers to grow the available varieties in the market. Capsicum, tomato and cucumber are grown with new technology of stem cutting, which can be the best alternate of vegetative propagation for the growers. The Department of Vegetable Science and Floriculture, CSK HPKV-Palampur is pioneer in developing vegetative propagation technology through stem cuttings in capsicum, tomato and cucumber. This technology can easily be accepted by the growers for multiplication of promising planting material.

16.2 BENEFITS

- Plants raised from cuttings produce early flowers, thus reduce the transformation period from vegetative to reproductive phase, hence, produce early crop.
- From a single plant number of rooted cuttings can be produced and hence will reduce the cost of the farmers in purchasing the costly seed.
- Polyhouse growers will be benefited more.
- It is easy to raise a healthy nursery of true to type plants.
- Saves the crucial time of the farmers, thus relive them from a stress of raising nursery.
- Cost of cultivation will be reduced, thereby increasing the profit margins as the seed of parthenocarpic varieties of cucumber is very high.

16.3 TECHNOLOGY

Cuttings are ready in about 2 weeks whereas propagation by seed takes 6 to 8 weeks. 10 cm long stem cuttings were taken from mother plant of capsicum, tomato and cucumber to clone varieties or regenerate more plants. One set of leaves is retained and pruned the remaining leaves by using a knife or scissors. Cutting should be tender and succulent. In case of tomato take cuttings from healthy indeterminate plants because indeterminate plants continue to grow whereas determinate plants grow to a certain size and set fruit. Cuttings taken from determinate tomato plants may root but may not grow or set blooms.

Mixture of soil, sand and FYM in the ratio of 1:1:1 was prepared and cuttings were planted in this mixture. Also cuttings were planted in plug trays containing soil less media of Cocopeat: Vermiculite: Perlite in a ratio of 3:1:1. After one week it was noticed that cuttings planted started rooting in the soil as well as in soil less media. After two weeks rooted cuttings were transplanted in the field.

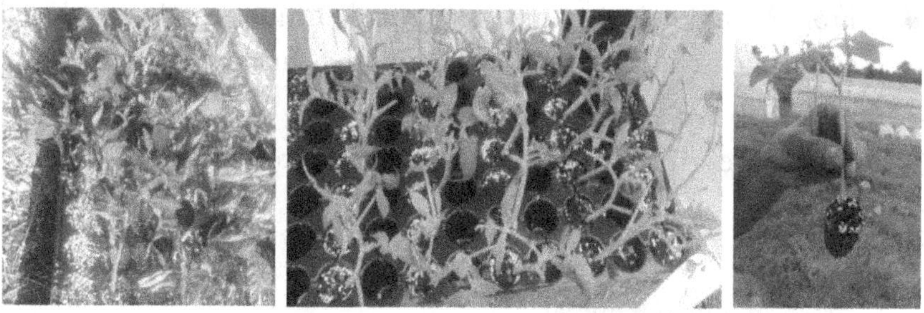

Capsicum, Tomato and Cucumber cuttings in rooting media

Rooting pattern in three potential crops

16.4 FIELD PERFORMANCE OF CUTTING RAISED PLANTS

The results obtained from the field performance under protected environment revealed that capsicum, tomato and cucumber plants raised through cuttings produced the same yield i.e. 2.0-2.5, 2.5-3.0 and 3.0-3.5kg per plant, respectively, as was recorded from plants raised through seeds. Significant observation recorded was early flowering and fruiting in all these crops.

Capsicum

Tomato

Cucumber

Research Outcome: (M.Sc. Thesis EFFECT OF ROOTING HORMONES ON STEM CUTTINGS OF TOMATO AND THEIR PERFORMANCE UNDER PROTECTED ENVIRONMENT

Vegetative propagation of tomato plant through stem cuttings has been proved an efficient method for cheaper multiplication and cultivation of tomato plants throughout the year. Growing hybrid/ improved cultivars of tomato through seed is expensive which a poor farmer cannot afford. So, minimization of cost incurred on the purchase of hybrid seed for every new crop will enable the tomato growers to get one extra crop of tomato through cuttings.

16.5 FUTURE PROSPECTS

With escalating prices of hybrid seeds of capsicum, tomato and parthenocarpic cucumber for protected cultivation, the major bottleneck in popularization of protected cultivation is the lack of suitable varieties or hybrids and their availability in the market. It has been observed that sometime farmers abandon their polyhouses due to non availability of quality planting material. The other important issue is the production of quality seedlings either through traditional method or hi-tech method of nursery production. Vegetative propagation is not common in vegetable crops in India and Department of Vegetable Science and Floriculture, CSK HPKV is pioneer in initiating research work on vegetative propagation technology of three important potential vegetable crops for protected cultivation. Vegetative propagation in capsicum, tomato and cucumber can be taken up by the farmers to avoid purchasing the costly seed of the desired hybrid again and again. Farmers can keep some plants as **'MOTHER PLANTS'** for taking cuttings and they can multiply the same hybrid on commercial scale for their use. For this purpose training, can be provided to farmers/ polyhouse growers so that an innovative technique of plant propagation can be demonstrated to them and problem of unavailability of quality planting material for polyhouse growers can be addressed.

CHAPTER - 17

HISTORY OF POMATO

Pomato i.e. tomato on potato rootstock is a wonderful horticultural innovation by horticulturists. It may be a very useful plant for urban areas of the country. We have been working systematically on various aspects of pomato development, rootstock-scion compatibility and nutrient management. It has a long history of its inception which is being discussed in sequence as :

1. 1915- "Burbank" marked about his one of the finding with herbaceous plant such as tomato and potato.
2. 1930– "Oscar Soderholm" (in Worcester, Massachusetts) - conceptualized the idea of grafted tomato onto potato plant, known as the "tomapotato".
3. 1977– Concept of tomato grafted onto potato was initially developed by the Germany based institute "Max Planck Institute for Developmental Biology in Tübingen. But plant "neither produced tomato fruit nor tubers of potato (Reinhard, 2008).
4. 1994– "The Max Planck Institute for Plant Breeding Research in Koln" produced a plant with fruit (Reinhard, 2008).
5. 2005– La´zaro E.P. Peres et al., 2005:- published a research paper entitled "Grafting of tomato mutants onto potato rootstocks: An approach to study leaf-derived signaling on

tuberization". An investigation was conducted at Departamento de Cie^ncias Biolo´gicas, Escola Superior de Agricultura "Luiz de Queiroz", Universidade de Sa˜o Paulo, Av. Pa´dua Dias, 11 CP. 09 CEP 13418-900 Piracicaba, SP, Brazil.

6. 2007- Nguyen Thi Trang Nha (Nong Lam Agriculture-Forestry University, based in Thu Duc District, Ho Chi Minh City, Vietnam:- Nha, suggested that, cross breeds could be developed when the seedlings of tomato and potato are of twenty days old.
7. "It is necessary to maintain the equilibrium in nursery time of crossbreeds, to obtain balanced yield.
   a. Produce from the crossbreed could be harvest after 60 days with 5kg of tomato fruit compared to control (7kg) and potato (0.5kg) than normal potato (0.7kg) (Anonymous, 2012).
8. 2010:- The kiambu prison conducted the trial (as directed in the literature sited from China as to grow fruit and tuber on same plant), a graft pomato plant was successfully developed but, did not yield many fruit or tubers (Lubbock online news, 2002)
9. 2013- "Thompson and Morgan"- UK based horticultural mail company launched the TomTato plant and sold this pre-grafted plant branded with a name "TomTato".
10. Thompson and Morgan stated that their business was the first to sell this grafted plant, and that the concept came from director Paul Handsord fifteen years ago in the United States, when he saw a garden where this grafted plant had been planted (a potato under a tomato) 2013 by Hall.
11. The TomTato plant was cultivated in a specialist laboratory in the Netherlands before being shipped to the UK and nurtured

in a greenhouse until it was ready to be sold (Wilkes, 2013). The "TomTato" plant produces five hundred white potatoes and five hundred cherry tomatoes.

12. In the same month, "Andrew and Fiona Boylan" of the Incredible Edible nursery in New Zealand unveiled a "Double UP Potato Tom" (Jude, 2013).
13. 2013:- According to "Jude," a dual-purpose plant may be created by grafting a tomato plant scion onto a potato stem.
14. 2013- According to "Jabr" grafting can respond positively as to impart resistance for, fungi, bacteria and viruses, provide a strong trunk and attracts more diverse group of pollinator.
15. 2013- "Hall" witnessed that a plant which produces both potato and tomato crops / dual crop, designated as a "veg plot in a pot", has been launched in the UK.
16. 2013- "Jabr" advocated that, some gardener and farmers have developed pomato plants. Which, yields tomatoes on the top portion and potatoes underground. These two crops vary in appearance, but they both belong to same genus.
17. 2013- "Barter" as a contributor to BBC Gardener's World, held the statement that 'numerous plants were developed through grafting', but had issue with taste of the product, so as it could not be a valuable plant to Thompson and Morgan company.
18. 2013- 'Guru' advised that, pomato product i.e. tubers of potato must be ready to harvest in 12th weeks in summer months and it could be ready to harvest as soon as leaves of potato begins to die, usually in early autumn.
19. 2013- "Greene" conferring that, a new wonder plant known as TomTato sold in the market, produces tomato fruits (cherry tomato) above the ground portion on vines, whereas, potato tubers of white colour in the soil and both the product were harvested from same plant (considered as an ideal plant).

20. Coxworth, 2013- Dual crop plant "pomato" is difficult to attain as it requires stem of the crop having same thickness to develop grafts and grafting is highly skilled process. In pomato plant potato foliage is not allowed to grow further, this plant only last for one season by the time potato tuber can be dig up and harvest the tomato fruits, when ready for picking (BBC News, 2013).
21. The Oregon Seed Company reported in 2014 that the grafted plant named "Tomtato" was developed in the United Kingdom. Ever since potatoes and tomatoes are closely related to each other, so they graft well collectively. It's not genetically engineered crop plant. Gardeners may now harvest a double crop of red cherry tomatoes and white potatoes from the same "TomTato" plant. (CBS Seattle Newsletter, 2014).
22. Springvale Garden Centre, 2014:- stated that, tomato and potato plants are naturally compatible with each other, and the idea of grafted tomato onto potato plant is not new idea, it has been done traditionally in 1900's but not commercialized before. This technique is considered to be best utilized as space saving, particularly for small gardeners.
23. Handsord, 2015:- conferring that, tomato and potatoes belong to same family and naturally compatible with each other. TomTato plant is produced by manual grafting with double cropping feature; crop is eco-friendly and safe as, there is no genetic modification.
24. The Orange County Register, 2015, showed that tomato and potato both the crop are the part of same genus / nightshade family i.e. tomato is of the *S. lycopersicum* and potato is of *S. tuberosum*. This plant also named as Ketchup 'n' Fries plant and could be popularized for home gardening especially in

United States. Nevertheless as a chimera-like twofer; Ketchup 'n' Fries are gathering the attention of more than just tomato gardeners.

25. Fresh Fruit Portal newsletter, 2015:- A Kenya based farm has produced grafted tomato onto potato plant could be utilized to maximize the use of small lands and this work was started after the trial was conducted in Kenya's Kiambu Prison Farm.
26. Mail online news, 2015:- shows that it is necessary to maintain the temperature variation of both the crop, as to ensure the perfect joining of two plants. A numeral of novelties that permits for gardening in small spaces, together with a system that produces vertically-stacked veggies in window, a ferris wheel-like mechanism, a mat that displays where to plant specially-prepared seeds.
27. Fresh Fruit portal newsletter, 2015; Business Daily, 2015:- pomato plant contributes to upsurge the food production more efficiently, from small land in urban environmental such as balcony could be utilized to get more amount of crop from same plant. Moreover on that, more beneficial for developing countries, as save on space, time and labour but not on cost of compromising with quality of produce. In addition to this, Grafting ensures resistance for kind of diseases such as, fungi, bacterial and viral, which helps to attract the pollinators and givers vigorous trunk to the delicate ornamental plants.
28. 2015:- Systematic work was started at Department of Vegetable Science and Floriculture, Chaudhary Sarwan Kumar Himachal Pradesh Krishi Vishvavidyalaya, Palampur, India.
29. 2015:- Jahanara Pervin (M.Sc. in Genetics and Plant Breeding), completed his thesis on entitled topic "Cell Compatibility Analysis of Pomato" at Sher-e-Bangla Agricultural University, Dhaka.

30. Pardeep Kumar and his co-workers released a study on the influence of grafting procedures on horticultural and quality attributes in tomato scion grafted on potato rootstock in 2016, and the inquiry has begun at the Chaudhary Sarwan Kumar Himachal Pradesh Krishi Vishvavidyalaya in Palampur, India.
31. 2016:- Viplove Negi (M.Sc. In Vegetable Science), completed his thesis on entitled topic "Studies on Development of Pomato through Grafting under Protected Environment" in Department of Vegetable Science and Floriculture, Chaudhary Sarwan Kumar Himachal Pradesh Krishi Vishvavidyalaya, Palampur, India.
32. 2016: Viplove Negi et al. released "Graft compatibility investigations in interspecific tomato-potato grafts" as a research article, at Chaudhary Sarwan Kumar Himachal Pradesh Krishi Vishvavidyalaya, Palampur, India.
33. 2017. Viplove Negi et al., 2017: Published a study report on how grafting tomato cultivars on potato rootstocks for increased returns affects horticultural and yield related variables in Department of Vegetable Science and Floriculture, Chaudhary Sarwan Kumar Himachal Pradesh Krishi Vishvavidyalaya, Palampur, India.
34. Guanghai Zhang and Huachun Guo released a study titled "Effects of tomato and potato heterografting on photosynthesis, quality, and yield of grafted parents" on February 14, 2018. Yunnan Agricultural University, Kunming 650201, China, Root and Tuber Crops Research Institute
35. 2019: Arefin et al., 2019: Published a research paper on 'Evaluation of Compatibility, Growth Characteristics, and Yield of Tomato Grafted on Potato ('Pomato')'. The investigation was carried out at the Department of Genetics and Plant

Breeding, Sher-e-Bangla Agricultural University, Dhaka, Bangladesh.

36. Siddikul Islam and his colleagues submitted a research article titled "Pomato: Double Harvest from a Single Plant" on March, 2019. The investigation took place in Majhian, West Bengal, India, at the "Dakshin Dinajpur KVK, Uttar Banga Krishi Viswavidyalaya."
37. Kumar et al., 2021. Completed a thesis entitled "Nutrient Management studies in Pomato under protected condition" under supervision of Dr. Pradeep Kumar at CSKHPKV-Palampur and recommended that 75% RDF - Fertigation at 4.56 gm^2 followed by 125% RDF + fertigation at 2.28 gm^2 found effective to increase the growth, yield and quality of Pomato.
38. Vandna et al., 2022. Completed her thesis work on "Rootstock-scion compatibility studies in Pomato" at CSKHPKV-Palampur.

CHAPTER - 18

'POMATO' OR 'TOMTATO': THE FUTURE RECOMBINANT HORTICULTURAL WONDER PLANT

18.1 WHAT IS POMATO...?

Pomato is a recombinant plant created by grafting tomato scions onto potato rootstocks, resulting in tomatoes on the surface and potatoes below. It possesses the qualities of both of its parents, since it is resistant to cold, like the potato, and heat, like the tomato. It may be cultivated both indoors and out in the fields. To achieve a distinctive double-cropping trait, each plant is grafted by hand. So it's not a genetic engineering breakthrough, but simply a straightforward graft of one plant's top onto the other. The potato and tomato plants have a tight genetic link, which is why this recombinant plant works.

Green Revolution occurred during the 1960s in India to achieve self-sufficiency in food grain production. On contrary, the present situation is entirely different as the major aim is to increase agricultural productivity from limited land resources by exploiting new innovative and productive techniques. With a sharp rise in population, urbanization, and industrialization, the cultivated land is shrinking day by day at an alarming pace. Consequently, it will be difficult to fulfill the quality of food requirements by the growing population, with the available resources. Rapid urbanization and industrialization will exacerbate the need for more high-quality food, particularly high-value vegetables, as land, water, and other resources become scarce. Further and more rural regions are now becoming peri-urban as the metropolitan outskirts grow. Food pricing and environmental stressors in peri-urban regions will be severe difficulties due to strong demographic demand. Increased demand for fresh veggies will result from quickly evolving eating patterns, increased health concerns, and rising affluence in metropolitan and peri-urban regions. To counter these challenges new technologies and inventions in vegetable grafting will play an important role.

Pomato has been viewed as a novel method for enhancing crop output per unit space, particularly in a tiny urban context such as a balcony or terrace garden. It appears as it came out of a genetic engineering lab, with tomatoes on one end and potatoes on the other. It is, however, a seedling tomato plant grafted on top of a potato plant, rather than a genetically modified miracle. There is no genetic modification; it is created through an all-natural and safe process. It sounds incredible, but it's true.

Because the two portions of the plant are genetically different and solely rely on each other for food and growth, this plant will not arise in nature and cannot be cultivated from seed. A simple tongue grafting procedure may be used to make a pomato or tomato plant by just leaving a few leaves of the tomato plant to nourish the potato plant's roots. In contrast to the former method, now a more common cleft grafting method is being utilized in many countries for producing pomato plants. In cleft grafting, the potato stems are given a horizontal cut sufficiently above the soil level, thereafter a vertical downward cut is made through the center of the stem to facilitate the insertion of the tomato scion, then a clear and smooth wedge-shaped cut is made at the base of the tomato shoots to be used as the scion, which must be of the same diameter as that of potato shoots to ensure the maximum possible contact of the cambium cell layers for the formation of a successful graft union and ultimately the joint is tied or held in place with the help of a sticky floral tape, polyethylene sheet or grafting clips to provide support to the grafted plant and of course to prevent moisture loss from the wounds.

Pomato plant at initial stage

Pomato plants at fruiting stage

The tomato scions are chosen for their fruits, while the potato rootstock serves as a solid and robust root system. After about 12-15 weeks of grafting over the summer months, the tomatoes are harvestable, and the potatoes may be picked once the tomato leaves begin to fall back, which is usually in early October. In India, potatoes are picked first, followed by tomatoes, which are allowed to mature further. The double species works because potato and tomato plants are strongly linked and belong to the same plant family, Solanaceae, with the same basic chromosome number of twelve.

Although the tomato was previously listed in a separate genus of *Lycopersicon* (*i.e. Lycopersicon esculentum* Mill.). In recent years it had been re-classified as *Solanum lycopersicum* L., the same genus as potato (*Solanum tuberosum* L.) indicative of their close relationship. But rather than just chromosome count, it is the compatibility of the graft union, where the all the important cambium cells found under the skin of the potato and tomato shoots need to match up for the graft to work.

Many research workers in the past had tried to develop this unique horticultural wonder plant, but due to the lack of any systematic research work, technical know-how, and methodology, it could not become a successful crop. In September 2013, the horticultural mail-order firm 'Thompson & Morgan' debuted grafted pomato plants in the United Kingdom, offering the pre-grafted plants branded as the "TomTato." In the same month, the 'Incredible Edibles Nursery' in New Zealand released a "Double UP: Potato Tom" for the same concept. According to Thompson & Morgan, this is the first time the plant has been professionally manufactured, and director Paul Hansord said how the concept for TomTato came about 15 years ago in the United States when he was visiting a garden where someone had planted a potato under a tomato as a joke. Grafting is a challenging technique since the tomato and potato stems must be the same thickness, and the study was required before commercialization to develop potato and tomato types that complement each other effectively. As a result, Thompson & Morgan put the Tomato facility through its paces before selling it. Plant production and grafting begin at a specialized laboratory in the Netherlands, before being brought back to the United Kingdom and grown in greenhouses until ready to sell.

Grafting provides disease resistance when the rootstock is robust, vigorous, and tolerant of abiotic conditions, such as heavy soils, but which would not otherwise produce as prolific or tasty fruits as a desirable cultivar. Growing Pomato plants saves farmers room, time, and labor without compromising the quality of their crop, which might have a big impact on nations like India. Grafting can also increase resistance to germs, viruses, and fungus, as well as attract a wider range of pollinators and offer a strong trunk for the fragile plants. Resistance to major soil-borne diseases is provided by resistant inter-specific rootstocks, and potato is one of the rootstocks used in many regions of the world. Grafting will be important in low-input sustainable horticulture in the future since it increases disease resistance and vitality in the crops.

18.2 WORK DONE AT CSK HPKV, PALAMPUR

In Himachal Pradesh, Department of Vegetable Science and Floriculture, CSK HPKV, Palampur has initiated systematic research work on development and popularization of this double-cropping plant. Initial experiments have indicated that if this crop is raised under protected environment, it can help in increasing the returns per unit area. To develop a Pomato or Tomtato plant a simple grafting technique is followed. The potato rootstock is sown approximately one month prior and after that either simple cleft grafting or tongue grafting is followed. The potato tubers are harvested when they mature and tomato plant is allowed to grow further.

- Pardeep et al., 2016, experimented on tomato grafted on potato to analyse the impact of grafting on fruit and tuber quality of both the crops, the outcomes revealed that no significant effect has been seen on the different horticultural parameters of grafting methods, but cleft grafting and tongue grafting performed better as the survival rate, yield and quality were enhanced in the combination of tomato scion "Avtar-7711 & GS-600" grafted on to potato rootstock "K. Himalini, K. Giriraj and K. Himsona."

- In the year 2016, Negi and his co-workers, elucidated that the maximum success rate in inter specific grafting of tomato-potato has been found with cleft grafting along with the combination of GS-600 + K. Himalini.
- In 2017, another research paper published by Negi and his coworkers mentioned that the pomato crop found to be more beneficial plant as the cost benefit ratio of pomato plants (1:2.12) was much higher than the solo crop of tomato and potato.
- An another achievement has been enlisted in tomato crops with respect to tomato grafting, as to popularize the pomato plant, fertilizer and fertigation treatment has become a necessity, because the pomato plant is combination of two crops and the need of nutrient doubled, so to standardize the fertilizers dose for pomato, Kumar et al., 2021 reported that 75% RDF= fertigation at 4.56 g m^{-2} followed by 125% RDF + Fertigation at 2.28 g m^{-2} found effective to increase the growth, yield and quality of pomato.

18.3 CONCLUSION

Finally, grafting a tomato plant onto a potato plant yielded fruits on the aerial sections and tubers on the subsurface area from a single tomato plant. It has the potential to alter the timing and quality of fruit harvests. It will also create opportunities for the amateur growers to develop new crop plant series. The development of this unique horticultural wonder plant may bring desirable changes in the present scenario of protected cultivation.

Luxurious Pomato Crop at CSKHPKV-Palampur

CHAPTER - 19

GRAFTING MIRACLE BRIMATO

19.1 INTRODUCTION

BRIMATO, known as egg and chips plant is a ground breaking recombinant plant developed through grafting approach by using potato as rootstock and brinjal as scion, which produce brinjal on upper portion and potato tuber in roots. This plant endures the properties of both the plant, which makes this plant popularized as dual cropping and space saving plant. It can tolerate or resist against biotic and abiotic stresses like soil borne diseases, nematodes, high and low temperature same as potato and brinjal. It can be grown indoor and outdoor likewise. It is a non-GMO plant which is developed through simple grafting technique. The reason for the successful graft union of brinjal and potato is that both the crops shares close genetic relationship as it shares same basic chromosome number and belong to the same family.

Brimato plant is the compatible combination of two crop plants brinjal and potato used as scion and rootstock respectively, which assured the double yield benefits from a single plant and can be considered as useful technology for utilizing the vertical space for future prospects as horizontal land expansion for agriculture is the most difficult objective to fulfill in the future. It is a combination of brinjal grafted on potato plant. As now a days breeding and genetic engineering has been seen as a most prominent technologies to resolve the problems regarding qualitative and quantitative traits of the plant. But vegetable grafting is one of the most innovative idea, can be used as eco-friendly and has safe process.

19.2 TECHNOLOGY

Brinjal and potato plants are closely related to each other as it belongs to same family, Solanaceae, sharing the same basic chromosome number 12. It is a plant which cannot be grown through seeds or other means in nature, because it's a combination of two genetically different parts of the plant, grafted together for obtaining two crops on a single plant. Healthy seedlings are important for obtaining successful grafted plants. This plant can be developed by tongue grafting or either by using cleft grafting.

In tongue grafting, firstly make a slant cut with the help of sharp knife on both rootstock and scion and then in between the slant cut, create a flat oval cut face of one to two inches deeper cut which provide better support to grafts and after that tie it with either by grafting clips or sticky floral tape. Cleft grafting is the approaches which can also be used for producing the grafted plant of Brimato. In this approach firstly horizontal cut on stem is given just above the soil thereafter in a centre of the stem a vertical cut in a downward position is given to facilitate the insertion of the brinjal scion. After that on the base of the brinjal scion wedge shaped cut is made, the diameter of brinjal and potato root should be same, so it can ensure the successful graft union establishment by increasing the chances of cambium cell layer establishment between both rootstock and scion. Grafts union are joined or tied either by using polythene sheet, sticky tape or with grafting clips, which helps plant to provide support and prevents the moisture loss from adjoining area to cure from the wound formation as well. In grafting, the rootstock plays important role because only healthy and vigorous rootstock can ensures the maximum possibility of successful grafting, while scions are selected for the fruit quality purpose. Brinjal fruit is ready to harvest after 75-80 days, whereas potato tubers are harvested firstly and brinjal is allowed to grow further, because it gives fruits for a long period of time.

19.3 BACKGROUND OF TECHNOLOGY

Since a very long period of time, work on vegetable grafting is going on throughout the world but due to lack of knowledge about the implementation of the grafting in vegetable crops, this technology in vegetable is not much popularized among the farmer. In recent period of time, Seed and Plant Company in England of Thompson and Morgan launched Pomato plant popularly known as horticultural wonder plant in 2013, and then in this series of grafted plant they introduced a new grafted plant which is known as Egg and Chips plant or Brimato. This company is selling the grafted plant with the name Egg and Chips plant. This plant named as egg and chips plant after a famous English dish of fried eggs and French fries, where egg stands for eggplant and chips as plant produce Potato. Thompson and Morgan claims that every egg and chips plant produces about 3-4 brinjal fruit and approximately 2 kg of potato tubers. For having additional benefits less hardy aubergine plant grafted on the tougher potato roots, it helps the plant to thrive in different environmental conditions. This plant reaches up to the height about 20 inches and flowering occurs in between July and August month. The best quality of this plant is that it can be grown as indoor and outdoor plant either in greenhouse or in a sunny sheltered position.

Brimato or Eggs and Chips Plant Developed by Thompson and Morgan Company

19.4 BENEFITS OF THE PLANT

Brimato plant is not developed through genetic engineering technique, it's a non- GMO plant, which is an environment friendly plant and relies on a grafting approach. It's a dual cropping, space saving and versatile plant, ideal for growing in a kitchen garden and small balcony and for seasonal vegetable growers. In a recent time, the breeding approaches and genetic engineering are commercially utilized in plants for many aspects, but these approaches are time consuming. Whereas, grafting emerges as convenient, easy to apply approach for solving the problems related to biotic and abiotic stress aspects. It helps the plant to carry disease resistance and tolerance to environmental stresses. Rootstock with vigorous roots system selection and scion with better fruit quality is the best combination for obtaining healthy crops with maximum returns per unit area from a single plant.

This technology could have momentous impacts in developing countries like India, where each vegetable grower with small land holding can also maximize productivity and income returns by saving time, utilizing space in a better way and by reducing the input cost in terms of labour and other practices. In addition to this, it provides resistance from soil-borne pathogens, bacterial, fungal diseases from nematodes, attract more diverse pollinator groups, and can also provide the robust stem support to delicate plant. Potato is one of the crops which is exploited as rootstock throughout the world, especially to encounter the soil-borne diseases. This technology can be utilized as a tool in order to upsurge the sustainable horticulture industrial income and area.

19.5 RELEVANCE OF BRIMATO IN INDIAN CONTEXT

Population in India is increasing at alarming rate which will create food crisis in coming years. Demographic pressure has forced the country to industrialization and urbanization due to which arable land is decreasing. Technologies like Pomato and Brimato will be useful as 'Space- saving plants' yielding more crops per unit area. Moreover such ideotype plants can be grown indoor / balcony or in pots, thus will provide more crops and returns. This can be a technological intervention to reduce dependence on limited agricultural land of the country.

19.6 WORK DONE AT CSKHPKV PALAMPUR

CSKHPKV Palampur, Himachal Pradesh, Department of Vegetable Science and Floriculture is pioneer in initiating the initial research work in order to popularize dual crop plants which can be developed and grown successfully. Department of Vegetable Science and Floriculture, CSKHPKV Palampur has successfully grown 'Pomato a double crop plant and has already conducted research trials to popularize Pomato. Similarly, a simple grafting technique can develop another dual crop plant called as 'Brimato' or Egg and

Chips plant. The sowing of potato rootstock and nursery of brinjal can be done approximately one month prior to grafting, and after that brinjal scion grafted on potato rootstock either by tongue or cleft grafting, use the scion and rootstock of same diameter, so it can ensures the maximum possibilities of successful compatibility of cambium cell layers in graft union. Potato tubers are harvested when it gains the marketable stage and the brinjal plant is allowed to grow further as it bears fruit for long period of time. Brimato plant yield brinjal fruit 400-500 gm/ plant and potato tubers 300-400 gm/ plant respectively.

Grafted Plant

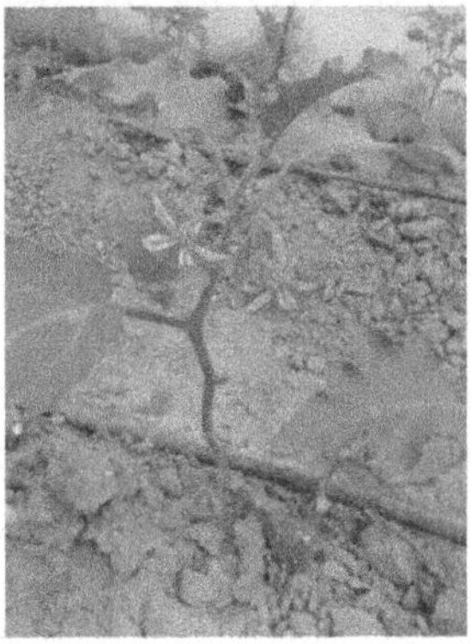

Flowering Stage

Fruiting Stage

BRIMATO PLANT DEVELOPED BY CSKHPKV-PALAMPUR

19.7 SUMMARY

Brimato is one of those plants which can create number of opportunities for making the best use of the space and land as it is known as dual cropping plant. Vegetable growers who have small land holding they can utilize it in a best way to increase return per unit area. Thus, in future this grafting miracle plant may bring fortunate and profitable changes in the present scenario of vegetable production.

CHAPTER - 20

IMPACT OF GRAFTING ON FRUIT QUALITY OF VEGETABLE CROPS

20.1 WHAT IS QUALITY?

Fruit and vegetable quality is typically characterized as "a collection of traits feature of a product capable of satisfying the demands of the consumer and determining its worth" (Peri, 2004). The quality traits are inherent and can be quantified according to the aim on the basis of product perspectives. From consumer point of view the definition of the quality is that it should be less tangible and quantifiable mainly focused on the consumer satisfaction (Shewfelt, 1999). Whereas, the quality of any product can be decided on the basis of qualitative standards, which can majorly affects the marketable value of any kind of vegetables or fruits. So, these qualitative standards are divided into two broad categories that are internal and external traits, helps to decide the value of the crops; internal traits elaborates the texture, flavor and antioxidants and other nutritional aspects how so ever the external traits covers the shape, size, colour, ensures the absence of defective product and freshness. Another way of categorization of valuable crop product can be done on the basis of their structural and functional quality, where structural quality includes the inherent traits present in the product such as biochemical properties like ascorbic content and sugar content and the functional quality traits analyses on the basis of visual or sensory traits preferred by the consumer for example sweet and sour taste of the edible fruits. (Huyskens- Keil and Schreiner 2003; Gruda, 2005).

Grafted plants products quality assurance is one of the intense issues emerging among the consumers. All elements that contribute to the appearance of quality should be considered while dealing with the uncertainty of quality. So, there are the major bodies are targeted for the standardization of product quality i.e. grower, distributor and the most important is consumer.

20.2 ROOTSTOCK EFFECTS ON FRUIT QUALITY

20.2.1 Appearance

(a) **Fruit size:** Fruit size is directly correlated with the yield as bigger the size of fruit more will be the yield. In context to the grafted plants, ensures the increase in the yield for kind of rootstock and scion combination in various vegetable crops. To understand the yield enhancement process in the grafted plants, its needed to know about the functional process of the grafted plant, as grafted plants ensures the vigorous root system, which enhances the physiological and metabolic mechanism by improving the water up-take efficiency, nutrient availability and photosynthetic activity of plant and ultimately increase the productivity of the crops. Grafted plants of tomato found with the positive increment in the fruit mass as compare to the non- grafted plants as confirmed by numbers of researcher (Yetisir and Sari, 2003; Colla et al., 2006; Huitron et al., 2007; Cushman and Huan, 2008; Turhan et al., 2011; Moncada et al., 2013).

Table 1.1: Impact of grafting on fruit size/ mass in various grafted vegetable crops

Crop	Scion	Rootstock	Mass of fruit	Yield	References
Watermelon	-	-	+	+	Proietti et al., 2008; Cushman and Huan, 2008
Melon (*Cucumis melo var. indorous*)	Incas	RS841	+ (10 %)	+	(Verzera et al., 2014).
Cucumber	RS841	Strong Tosa	+	+	Goreta Ban et al., 2014
	Akito and Ekron	P 360, PS1313 and Affnye	+	+	Colla et al., 2012, 2013
Tomato	Lemance F_1 and Star Fighter	Beaufort	+	+	Pogonyi et al., 2004; Theodoropoulou et al., 2007
Cherry Tomato	Classy	Brigeor and Maxifort	Nil	Nil (No effect)	Schwarz et al., 2013
	Piccolino	Brigeor and Maxifort	+	+	Schwarz et al., 2013
Tomato	Rita	Beaufort, Heman, Joint, P-1614 and RS 1427	+	+	Romano et al., 2000

[Table Contd.

Contd. Table]

Crop	Scion	Rootstock	Mass of fruit	Yield	References
Aubergine	Black beauty	*S. torvum* Sw., *S. incanum* L. x *S. melongena*, *S. melongena* x *S. aethiopicum*	+ (Higher no. of fruits + mass)	+	Gisbert et al., 2011b
	Mission Bell F_1	PG-3, Beaufort, Energy (Tomato)	- (lower no. of fruits/ plant)	-	Leonardi and Giuffrida, 2006
Pepper	Coyote, Almuden	Charlot and Foc	- (11-21%)	-	Gisbert et al., 2010
Sweet Pepper	Triple Star	AR96029	+	+	Leal- Fernandez et al., 2013

(+ = higher/increase), (- = decreases)

(b) **Fruit shape:** Fruit shape is the ratio of the diameter of the fruit taken as equatorial and longitudinal directions. Melons are example where the shape of fruit of grafted plants remains unaffected as mentioned by Colla et al., 2006 a; Rouphael et al., 2008; Soteriou and Kyriacou, 2015.

(c) **Colour:** Fruit colour of the grafted watermelon pulp was influenced because grafting affects the lycopene content of the fruit and also limits the inconsistency among two variables as confirmed in certain grafted combinations. Delay in the ripening process of watermelon was reported when scion of watermelon grafted onto the interspecific *Cucurbita* hybrids rootstock. Whereas, in tomato not significant effect of grafting been seen on the colour of the fruit, but some slight variation in skin colour of tomato fruit was confirmed in grafted plants as per the results of Oztekin, 2009; Schwarz et al., 2013; Davis and Perkins-Veazie, 2015; Soteriou et al., 2014.

(d) **Exocarp attributes:** Grafting performed in watermelon affects the thickness of the fruit rind, and rind thickness is one of the commercial interest trait. As the rootstock minimum rind thickness was found in response to the rootstock than that of variety/ cultivar used as scion material. How so ever, in case of chilli pericarp, thickness was significantly influenced in the fruits of grafted plants (Alexopoules et al., 2007; Proietti et al., 2008; Donas-Ucles et al., 2014; Kyriacou and Soteriou, 2015).

(e) **Absence of defects:** Defects may be considered as any kind of disorder appears, in response to mineral deficiency or environmental conditions.

Table 1.2: Impact of grafting on fruit shape index in various grafted vegetable crops

Crop	Scion	Rootstock	Fruit shape index	References
Cucumber	Ekron	Affyne	+ (4% higher)	Colla et al., 2013
	Ekron	P360		
Tomato	-	Beaufort and Arnold	+	Turhan et al., 2011
Aubergine	Crystal F_1	*S. incanum* x *S. melongena* and *S melongena*	+ (length)	Gisbert et al., 2011a
	Black Bell	*S. torvum*	(4% longer fruits)	Casaniti et al., 2011

(+ = higher/increase)

Table 1.3: Impact of grafting on fruit colour in various grafted vegetable crops

Crop	Scion	Rootstock	Colour findings	References
Melon	*Cucumis melo* var. *cantaloupensis*	P360	L* and a/b*(Modified colour-CIELAB values)	Colla et al., 2006a
Cucumber	-	-	L* (4% higher) and a/b* (5% lower)- Hunter colour values	Colla et al., 2012
Tomato	-	-	L* (not affected)a/b*(lower)	Ulukapi and Onus, 2005
Aubergine	Birgah	*S. torvum*	Darker with less vivid	Moncada et al., 2013

(L*)= Brightness, (a/b*) =redness/yellowness ratio

Table 1.4: Impact of grafting on Exocarp attributes in various grafted vegetable crops

Crop	Scion	Rootstock	Exocarp thicknessRind/ Pericarp	References
Watermelon	Watermelon	*C. maxima* x C *moschata*, *Lagenaria siceraria* (Molina) Standl., *C. maxima* and *C. moschata* landraces	Significant	Alexopoules et al., 2007; Proietti et al., 2008; Kyriacou and Soteriou, 2015; Yetisir et al., 2003
	-do-	*C. pepo*	Non- significant	Davis and Perkins-Veazie, 2005
Pepper	Palermo	AR40	Significant (5% higher)	Donas-Ucles et al., 2014

Table 1.5: Impact of grafting on Absence of defects in various grafted vegetable crops

Crop	Scion	Rootstock	Appearance of Defects	References
Watermelon	Watermelon	*C. maxima* x C *moschata*, *Lagenaria siceraria* (Molina) Standl.	No	Jang et al., 2014
Melons		Shinotosa	Vitrescence (observed in 89% of melon)	
		Cucumis rootstock	Lower Vitrescence	
Tomato	Classy	Brigeor, Maxifort (2 light levels)	Reduce incidence of blossom end rot	Krumbein and Schwarz, 2013

20.2.2 Texture

Firmness of the fruit is one of the most desirable characters which could be affected or related to the concentration of calcium. Number of other mechanisms also showed their impact on the firmness such as transpiration, solubilization, cell wall protein and architecture and water relation etc (Saladie et al., 2007).

20.2.3 Sugar Content and Acidity

Several organoleptic compounds are responsible for the flavour and aroma of any kind of food product, these compounds includes amino acids, sugar compounds and numbers of volatile compounds. Sugar content and acidity are considered as chief parameters in order to enhance the aroma and flavour in several vegetable and fruits products (Kader, 2008).

20.2.4 Health- promoting substances

(a) **Carotenoids:** Carotenoids are responsible for the colour development on the fruits, such as lycopene, alpha & beta-carotenes in crops like watermelon, melons, tomato, brinjal and other vegetable crops.

(b) **Ascorbic acid:** Watermelon grafted on *C. maxima* x *C. moschata* rootstock increased total ascorbate concentration by 7.3 percent, owing mostly to a rise in the oxidized dehydroascorbate form (Proietti et al., 2008). Huang et al. (2009) demonstrated that grafting favorably improved vitamin C content in cucumber, as an increase in the ascorbic acid content of the fruits was obtained by grafting on to 'Fig leaf Gourd' and 'Chaofeng Kangshengwang' both under saline and non-saline nutritional solution conditions. Turhan et al. (2011) demonstrated that the

Table 2: Impact of grafting on Texture in various grafted vegetable crops

Crop	Scion	Rootstock	Pulp firmness	References
Watermelon	Watermelon	Interspecific hybrid (*C. maxima* x *C. moschata)*	Increase	Soteriou et al., 2014; Soteriou and Kyriacou, 2015)
	Diploid and triploid	(*C. maxima* x *C. moschata)*	Increase (7-36% in diploids and 14-55 % in triploids)	Soteriou and Kyriacou, 2015
	Watermelon	Cucurbit hybrid	Increase	Yetisir et al., 2003; Cushman and Huan, 2008; Bruton et al., 2009
Melons	Cyrano	P360	Increase (19 - 32%)	Colla et al., 2006a
inodorus melon	Homerunstar	Shintosa	Decrease	Jang et al., 2014
Cucumber	Kema or Kifujin New Type	Big Ban Kitora (Blossomless)	Increase	Morishita, 2001
Tomato	Big red	Heman , Primavera	no impact	Khah et al., 2006
Aubergine	Tsakoniki	*S. torvum, S, sisymbriifolium*	Negative impact	Arvanitoyannis et al., 2005

Table 3: Impact of grafting on Total soluble sugars (TSS), pH and Titrable acidity (TA) in various grafted vegetable crops

Crop	Scion	Rootstock	Sweetness	References
Watermelon	-	Pumpkin	Decreases	Soteriou and Kyriacou, 2015
	Diploid watermelon	Cushaw pumpkin (*C. argyrosperma*), Squash (*C. pepo)*	Decreases TSS	Davis and Perkins-Veazie, 2005
	-	*L. siceraria; C. maxima* x *C. moschata*	No impact on TSS	Proietti et al., 2008 ; Soteriou and Kyriacou, 2015
	-	*L. siceraria; C. maxima* x *C. moschata*	No impact on specific sugars (glucose, sucrose or fructose)	Proietti et al., 2008 ; Soteriou et al., 2014
Aubergine	Faselis, Pala	*S. torvum*	Decreases TSS	Khah, 2011
	Rima	Heman, Primavera	No impact on pH, TSS, TA	Khah, 2011

Table 4.1: Impact of grafting on carotenoids in various grafted vegetable crops

Crop	Scion	Rootstock	Carotenoids content	References
Watermelon	-	*C. maxima* x *C. moschata*	Increase (lycopene 10-40%)	Proietti et al., 2008 ; Soteriou et al., 2014; Kyriacou and Soteriou, 2015
	-	*L. siceraria*	Increase (lycopene 11-27%)	Candir et al., 2013
	Triploid watermelon	*C, argyrosperma, C. pepo*		Davis and Perkins-Veazie, 2005
Melons	-	AS10, Elsi, P360 and RS841	Increase Lutin content about 5.5 – 13.7 mg kg^{-1}), 8-fold beta carotene	Verzera et al., 2014
	-	P360	56% alpha and	
Tomato	Jeremy	Maxifort and Beaufort	Decrease lycopene	Miskovic et al., 2008
	Jeremy	Madonna (Aubergine)	Increases lycopene	
Bell pepper	Fascinato and Jeanette	Terrano	Increase beta- carotene and antioxidant activity	Chavez-Mendoza et al., 2013

Table 4.2: Impact of grafting on ascorbic acid in various grafted vegetable crops

Crop	Scion	Rootstock	Ascorbic Acid content	References
Crop	**Scion**	**Rootstock**		**References**
Watermelon	Watermelon	*C. maxima* x *C. moschata*	Increases (7.3%)	Proietti et al., 2008
Cucumber	-	Fig leaf and Chaofeng Kangshengwang	Increase	Huang et al., 2009
Tomato	-	Beaufort, Arnold	Increases	Turhan et al., 2011

Table 4.3: Impact of grafting on Phenolic compounds in various grafted vegetable crops

Crop	Scion	Rootstock	Carotenoids content	References
Tomato	Cherry tomato	Zarina	Increases (total flavonoids, hydroxycinnamic acids and rutin	Sanchez-Rodriguez et al., 2012
	Tamaris	Heman, Efiato and Maxifort	Decreases (total phenol)	Vinkovic- Vrcek et al., 2011
Aubergine	Cristal F_1	*S. macrocarpon* L.	Increases phenolic content	Gisbert et al., 2011b

ascorbic acid concentration varies depending on the scion-rootstock combination using tomato plants grafted on to 'Beaufort' rootstock, which had greater ascorbic acid content than the same plants grafted on to 'Arnold' rootstock.

(c) **Phenolic compounds:** Number of natural antioxidants are present in vegetable crops such as hydroxycinnamic acids, naringenin, chalcone and rutin in tomato (Slimestad et al., Sanchez Rodriguez et al., 2012).

20.3 CONCLUSION

The combination of genetic, environmental, cultural, and postharvest variables determines vegetable quality. In future, to determine the ideal combinations of these parameters to optimize vegetable quality will be a problem. Among the several aspects, the choice of appropriate rootstocks for plant grafting is vital. The selection of an adequate rootstock for a certain scion is thus critical not only for maximum output but also for fruit quality. As a result, grafting can be a significant strategy for improving fruit quality in addition to better managing soil-borne illnesses and improving abiotic stress tolerance.

Nevertheless, in certain scenarios, grafting could harm product quality; thus, appropriate cultural practices must be considered to minimize the negative effects that may result from the adoption of certain rootstocks expressing tolerance to specific stresses or determining a higher plant vigor. More emphasis should be placed on novel species that will be grafted in the future (e.g. artichoke). For the successful selection of acceptable rootstocks and the production of high-quality fruits from grafted vegetable crops, an integrated strategy based on collaboration among quality specialists, agronomists, physiologists, and breeders is required.

CHAPTER - 21

ENACTMENT OF GRAFTING IN UNDERUTILIZED CROPS

Though vegetable grafting is used for providing resistance to various stresses through rootstocks in fruit vegetables only, but there are other group of vegetables where it can be performed with different objectives. Commercially only tomato, brinjal, cucumber and bell-pepper are being grafted worldwide for countering major biotic as well as abiotic stresses. We don't have any experience for grafting these underutilized vegetables, but for academic purposes as well as to new researchers it may be of interest. Literature provide some information on grafting of okra plants, globe artichoke and green beans. The available information for artichoke and green bean is discussed as below:

21.1 GLOBE ARTICHOKE

Globe Artichoke a known mediterranean perennial vegetable belongs to compositaeae family, comprises of two cultivated species namely Cardoon (*C. cardunculus* var. *altilis* DC) and globe artichoke (*C. cardunculus* var. *scolymus* L.). The ancestor of the cultivated globe artichoke is wild cardoon/ artichoke thistle (*C. cardunculus* var. *sylvestris* (Lam.) Flori) (Rouphael *et al.*, 2012; Colla *et al.*,

2013). In a recent period of time, vegetable grafting is mostly utilized in commercial fruit bearing vegetable crops of solanaceous and cucurbitaceous family such as in tomato, brinjal, chilli, bell pepper, watermelon, melons, cucumber and gourds etc. Whereas in current scenario, this technique has been also used with respect to other underutilized crops such as in globe artichoke (Ciccarese *et al.*, 2012; Trinchera *et al.*, 2013; Temperini *et al.*, 2013). Grafting is one of the most prominent agrotechnology which could be used to combat numerous environmental stresses, in the context of globe artichoke Verticillium wilt, is one of the major soil borne fungi i.e. *Verticillium dehliae* with worldwide spread and becoming a sensitive threat for artichoke production. In 2012, number of reports revealed the impact of grafting on Verticillium wilt as it leads to impose tolerance in artichoke plants (Cirulli *et al.*, 2010). Numerous studies have been reported for resistance for *V. dehliae* in artichoke, in a case study accompanied by Ciccarese *et al.*, 2012 examine collection of cardoon including wild and cultivated species for resistance to *Verticillium* wilt under controlled conditions by artificial inoculation of fungus. Results publicized that, out of forty four only nine exhibited highest percentage of resistance to wilt deprived of any disease symptoms. On other hand, the nine resistant rootstocks of cardoon could be utilized as more promising tool predominantly in those areas where the soil is ominously infested with Verticillium wilt. Likewise, seed propagated globe artichoke selection "T3" whichever non-grafted or grafted on to the "Bianco gigante inerme a foglia intera" cultivated cardoon, assessed to know the crop performance in infested soil with *Verticillium* spp. for three years and the outcomes confirms that higher yield on an average of 44-53% with grafted plants as compared to the non-grafted ones (Temperini *et al.*, 2013). In addition to this, in another report grafted plants of globe artichoke significantly enhanced the yield and yield components when cultivated on verticillium infested as well as non-infested soil with seed-propagated globe artichoke hybrid "Concerto" grafted on to cultivated variety of cardoon i.e. "Belgio" whereas, the incidence of Verticillium wilt was found lower in grafted as compared to non-grafted plants. In same investigation, researchers assessed the compatibility of two seed globe artichoke cultivars grafted onto cultivated cardoon cultivar i.e. "Romolo" and "Istar" onto "Bianco avorio" and "wild cardoon", respectively and meanwhile splice grafting was found most accurate method of grafting to achieve maximum success rate with respect to globe artichoke (Temperini *et al.*, 2013). Splice grafting method was found most suitable with 78% success rate, when two cultivars of globe artichoke were self-grafted, cross-grafted and grafted on the rootstock "Bianco avorio" with confirmed higher grafting survival rate ranging from 82% to 92% (Temperini *et al.*, 2013).

According to Edelstein *et at.*, 2004 compatibility between rootstock and scion are the vital aspect in vegetable grafting, which depends on the different variables of the crop such as genetical, anatomical and physiological. Trinchera *et al.*, 2013 conveyed that the degree of graft compatibility or incompatibility and their mechanism between scion and rootstock was concomitant with the healing time of the two discrete units between two living matter or plants/ bionts. During the another investigation, when the longitudinal sections of the grafted connections were imaged under the scanning electron microscopy after three days of grafting, number of interconnection structures, vascular re-arrangements and callus development in between two grafting components has been visualized during the formation of graft union, meanwhile "Romolo- Sardo" associated with wild Cardoon "Sardo" have ability to produce carbohydrates and pectic at graft interface (Trinchera *et al.*, 2013).

21.2 GREEN BEANS

Vegetable grafting is emerging as a most consistent technique to grow green beans in soil based culture and in hydroponics under green house, particularly to prevent the environmental stresses; it can be biotic and abiotic stresses and this agrotechnology has been used since 2010 in Portugal. Green beans are in the interest of grower because of ease in management practices and its intense ability of environmental sustainability, which makes it more preferable for the organic cultivation. Generally this crop has been grown under greenhouse twice in a year and needs the high rate application of fertilizers without crop rotation. Due to this heavy application of fertilizers and pesticides as well led to upsurge the chances of soil salinity and soil borne diseases in such kind of soil, whereas the most known disease such as root knot nematode and caused by fusarium wilt, which lessen the production of this crop in the areas where it is grown on large scale (Rodrigues, 2010).

Meanwhile, some species of green bean such as *Phaseolus vulgaris* and *P. coocineus* has been reported, with enhanced activites like nutrient uptake, tolerance to salinity and soil borne diseases and act as a natural resistant source. The species of green bean *P. coocineus* was introduced from Meso-America to Europe in the seventh century, whereas, this species was grown from thousands of years particularly in Meso-America region, while the centre of origin of this species is Highlands of Mexican region mainly cultivated at the verge of cultivated plots of bean and the floral morphology of this species that attain high degree of allogamy with marvelous breeding efficiency and have well developed tap-root system with tuber and secondary roots, which contains fair amount of starch (Hernández-Bermejo and León, 1992; Delgado-

Salinas, 1988; Giurca, 2009). In comparison of *P. coccineus* attains great vigour *than P. vulgaris* have weak root system. The rustic form of *P. coccineus* known for resistance against fungi, virus and bacteria that is *Colletotrichum lindemuthianum*, bean golden mosaic virus and *Pseudomonas syringae* Van Mall; *Xanthomonas phaseoli*, respectively (Delgado- Salinas, 1988; Giurca¢, 2009; Labuda, 2010). There is a cultivar with indeterminate growth habit, commonly grown in Portugal named as 7-year bean with special feature vegetative growth at crowns of root allow re-growth for long period of years. The best approach of grafting in green bean is splice grafting along with two training support, which can endures sets of two stems; however in non- grafted and selfed plant this can be practiced only once (Colla *et al.*, 2017). In a case study, Giurca in 2009 initated an investigation at two different sites with low input of soil mineral nutrients and the results showed that the scion "Oriente" grafted onto rootstock "7-year bean" could be utilized as smart strategy to upsurges the tolerance against *Fusarium oxysporum* f. sp. *phaseoli* commonly known as vascular wilt disease, due to improved root system and nutrient absorption capacity of the roots as compared to common bean cultivars.

Even so, when scion "Oriente" and "Vagem Rajada" grafted onto rootstock of "White Emergo" and "7-year bean" denoted as P2 and P3 respond significantly well at both the experiment sites and found with increased nutrient content particularly for P and at second site only for Ca content as compared to self and non-grafted ones. Nevertheless, in another finding author's confirmed the two resistant quantitative trait loci associated for vascular wilt disease species (*Fusarium oxysporum* f. sp. *phaseoli)* in common bean (*P. vulgaris)* and showed about nine to twenty two per cent of the total genetic variation (Schneider*et al.*, 2001; Hagerty *et al.*,

2015; Mourao *et al.*, 2016). Some investigators elucidate that the resistance for *Meloidogyne javanica* could be achieved only if the resistant rootstock of *Phaseolus vulgaris* is used instead of susceptible rootstock and irrespective of scion section (Mourao *et al.*, 2016).

CHAPTER - 22

RESEARCH BACKGROUND IN VEGETABLE GRAFTING: A CASE STUDY OF CSKHPKV-PALAMPUR

Systematic research work was initiated in 2010 in the Department of Vegetable Science and Floriculture, CSKHPKV, Palampur and various experiments were conducted to screen resistant rootstocks for bacterial wilt, nematodes and gummy stem blight. Researchers from their thesis on different aspects of vegetable grafting reported significant and interesting findings. Some of the research findings are being discussed here for the benefit of researchers, stakeholders and students as below:

1. **Evaluation of rootstocks for bacterial wilt tolerance in bell pepper (*Capsicum annuum* L. var. *grossum* Sendt) under protected conditions** (Rana et al, 2015)
   - **Major findings:** The present investigation was carried out at Palampur during 2014 using five rootstocks of chilli and three of brinjal plus one control (non grafted plots of hybrid Indra). Commercial hybrid 'Indra' was used as a scion. Weekly record on bacterial wilt incidence showed that only non-grafted (control) plants i.e. Indra showed wilting symptoms while all the rootstocks were resistant to bacterial

wilt. Less survival rate of control /non grafted plants was due to the incidence of bacterial wilt. Therefore, from this study it may be inferred that grafting can be an effective strategy to manage bacterial wilt incidence in bell-pepper.

2. **Grafting tomato on potato rootstocks and its effect on quality traits (Kumar et al, 2016)**

   - **Major Findings:** The study was conducted in a naturally ventilated polyhouse during September 2015 to May 2016. Two indeterminate tomato hybrids [Avtar-7711 (S1) and GS- 600 (S2)] were used as scions and three potato cultivars as rootstocks [Kufri Himsona (V1), Kufri Himalini (V2) and Kufri Giriraj (V3)]. Treatment combination V1G2S2 (V1=Kufri Himsona, G2=Cleft grafting and S2=GS-600) and five other treatments recorded significantly higher TSS in potato tubers than non-grafted (control) plants. All graft combinations recorded significantly higher starch content in potato tubers than from non grafted (control) plants. The highest starch content (24.58%) was recorded in treatment combination V1G1S2 (V1=Kufri Himsona, G1=Tongue grafting and S2=GS-600). Maximum TSS (7.60°Brix) in tomato fruits was recorded in treatment combination V3G1S1 (V3 Kufri Giriraj, G1= Tongue grafting and S1= Avtar 7711).

3. **Evaluation of Chilli and Brinjal rootstocks for growth, yield and quality of bell pepper (Capsicum annuum L. var. grossum Sendt) under protected conditions (Kumar et al, 2016)**

   - **Major Findings:** The chilli rootstock PI-201232 was the most suitable rootstock for bell pepper. The chilli rootstocks PI-201232, AVPP0205, VI-037556, Surajmukhi and Pant C-1 were found to be more suitable for bell pepper in comparison to brinjal rootstocks. It is concluded that chilli rootstocks were found more compatible over brinjal rootstocks for bell pepper scion variety Indra. Use of rootstocks can be an alternative approach to combat wilt, increase disease tolerance and vigour in bell pepper.

4. **Effect of grafting techniques on horticultural and quality traits in tomato scions grafted on potato rootstocks (Kumar et al, 2016)**
   - **Major Findings:** Tomato hybrids Avtar-7711 and GS-600 were used as scions and potato cultivars Kufri Himsona, Kufri Himalini and Kufri Giriraj were used as rootstocks. Tomato scions were grafted on potato rootstocks using tongue grafting and cleft grafting methods. Cleft grafting is reported as best grafting method exhibiting maximum survival of pomato plants (84.13%) and a high degree of grafting success rate (72.95%). For yield and quality traits the effects of tongue grafting was more pronounced.
5. **Horticultural and yield related traits as influenced by grafting tomato cultivars on potato rootstocks for higher returns (Negi et al, 2017)**
   - **Major Findings:** Pomato is an innovative plant developed by simple grafting of tomato as a scion on potato as a rootstock. Potato and tomato are the crops of high economic importance and being genetically similar they can be easily grafted together and grown as a single plant producing potatoes and tomatoes on the same plant. With rapid urbanization and industrialization cultivable land is becoming the most limiting resource especially in urban areas making intensive cultivation practices inevitable to produce more crops per unit area to fulfill the food quality requirements of the ever increasing population. Crop plants

like pomato can be a suitable alternate in maximizing the potential of crops with efficient utilization of land and other resources. No systematic research work has been conducted in India on this aspect, therefore to generate substantial practical and useful information the present investigation entitled "Studies on development of pomato through grafting under protected environment" was undertaken at Vegetable Research Farm, Department of Vegetable Science and Floriculture, CSK HPKV, Palampur, Himachal Pradesh with the objectives to develop pomato plant through grafting, to find out the best recombinant on the basis of yield and component traits and to work out the cost of cultivation. The experiment was laid out in a Randomized Block Design with three replications and data was recorded on various horticultural and quality traits for potato and tomato. Kufri Himalini has been found to be the best rootstock for tomato scion. Cleft grafting is reported as best grafting method which resulted in maximum survival of grafted plants as well as maximum grafting success rate. Whereas, the effect of tongue grafting was more pronounced on yield and quality traits of either of the crop species. GS-600 was found to be the best compatible scion for potato rootstocks. Rootstock-scion combination of Kufri Himalini and GS-600 was found to be the best recombinant on the basis of yield and component traits. Treatment combination V2G1S2 (V2 = Kufri Himalini, G1 = Tongue grafting and S2 = GS-600) resulted into maximum survival of grafted plants (93.25%), tuber yield per plant (471.80 g), tuber yield per square meter (2.83 kg) in potato and number of marketable fruits per plant (22.55), marketable fruit yield per plant (1.74 kg), marketable fruit yield per square meter (10.44 kg) and harvest duration (97.33 days) in tomato. Minimum days to first flowering (47.33), days to first harvest (140.00) and maximum ascorbic acid content (21.34mg/100g) in tomato was recorded in treatment combination V3G1S2 (V3 = Kufri Giriraj, G1 = Tongue grafting and S2 = GS-600). Maximum TSS (6.47°Brix) in potato, average fruit weight (89.32 g) and plant height (276.34 cm) in tomato were recorded in treatment V1G2S2 (V1 = Kufri Himsona, G2 = Cleft grafting and S2 = GS-600). Treatment combination V2G2S1 (V2 = Kufri Himalini, G2 = Cleft grafting and S1 = Avtar (7711)) recorded maximum fruit length (5.42 cm) and fruit width (5.72 cm) in tomato. Maximum average tuber weight (139.04 g) in potato and TSS (7.60°Brix) in tomato fruits were recorded in treatment V3G1S1 (V3 = Kufri Giriraj, G1 = Tongue grafting and S1 = Avtar (7711)). Maximum pericarp thickness (8.23 mm) in tomato was recorded in treatment combination V1G1S1 (V1 = Kufri Himsona, G1 = Tongue grafting and S1 = Avtar (7711)). Treatment combination V1G1S2 (V1 = Kufri Himsona, G1 = Tongue grafting and S2 = GS-600) recorded maximum starch content (24.58%) in potato. Maximum number of tubers per plant (4.81) in potato were recorded in treatment combination V2G1S1 (V2 = Kufri Himalini, G1 = Tongue grafting and S1 = Avtar (7711)). Data indicated that there was no significant effect of interaction between rootstocks, grafting methods and scions on days to harvest in potato. Pomato resulted in cost-benefit ratio of 1:2.12 which was significantly higher than sole cultivation of tomato (1:1.93) and potato (1:0.26) under protected environments.

6. **Effect of grafting tomato on different rootstocks under protected environment in mid hills of NW Himalayan region (Kumar et al, 2017)**
   - **Major Findings:** Grafting plays an important role in increasing the growth, yield and quality of the tomato. The study also indicated that different rootstocks affected the quality of the fruits. Grafting has, therefore, a great potential to increase the yield and growth of tomato.
7. **Vegetable grafting: a boon to vegetable growers to combat biotic and abiotic stresses (Kumar et al, 2015)**
   - **Major Findings:** Grafting provides a site specific management tool for soil borne diseases. This technique is considered eco-friendly for sustainable vegetable production because the resistant rootstock reduces dependence on agrochemicals. Grafting improves quality of the plant and is used to induce resistance against low and high temperature. Since grafting gives increased disease tolerance and vigour to crops, it will be useful in the low input sustainable horticulture of the future.
8. **Standardization of Robotic Grafting in Bell-Pepper (*Capsicum annuum L. Var. Grossum* Sendt.). (2020)**
   - **Major findings:** For robotic grafting the age of scion and stock should be same and about 45 days old age. Maximum yield, quality and other horticultural traits were superior in plants when both scion and stock were 45 days old. Plants grafted on rootstock PI-201232 produced maximum no. of fruits, produced higher yield and quality as compared to other rootstocks. Treatment RS3A3 (RS3 = PI-201232 and A3 = 45 days old scion and rootstock) was found to be better than other treatment in almost all traits. Hence,

it may be concluded that robotic grafting is an effective technique to improve various horticultural and quality traits thereby, increasing total and early yield, increase tolerance to various soil borne diseases, improve vigour of plants and higher grafting success rate as compared to manual grafting.

9. **Rootstocks and Grafting Interaction Studies for Gummy Stem Blight (*Didymella bryoniae*) and Horticultural Traits in Parthenocarpic Cucumber. (2021)**
   - **Major findings:** The investigation confirms that the rootstocks and grafting techniques play a remarkable role in affecting growth and development of the cucumber. Rootstock and scion compatibility is crucial for getting desired results in grafted plants. Local varieties of different cucurbits were used for grafting a popular parthenocarpic cucumber hybrid Kian. Bottle gourd rootstock was found as one of the best rootstocks for grafting and it recorded higher plant growth and yield. Evaluation of characters based on grafting techniques revealed that splice grafting technique was the easiest method to be utilized with positive significant outcomes. Therefore, on the basis of outcome obtained from the analysis of various growth and yield attributes, resulted that bottle gourd is best compatible rootstock along with splice grafting technique for securing high yield under protected conditions.

10. **Effect of Fertilizers and Fertigation Treatments on Pomato Growth and Yield under Protected Environments.** ***(2021)***
    - **Major Findings:** Pomato is said to be the amalgamation of potato and tomato, produced by grafting, where potato is used as a rootstock and tomato as a scion. The experiment was conducted under controlled conditions at

Vegetable Research Farm, Himachal Pradesh Agricultural University, Palampur to investigate the effect of fertilizers on horticultural traits of pomato plant. The experimental material consisted of fourteen different fertilizer treatments comprising three controls T1 (100% RDF + FYM), T2 (100% RDF Sole Tomato) and T3 (100% RDF Sole Potato) for which potato cultivar "Kufri Himalini" and tomato cultivar "Palam Tomato Hybrid-1", served as grafting parents. Most of the traits under investigation were found to be significantly affected by different fertilizer treatments. The effect of fertilizer treatments on survival of grafted plants was found to be non-significant. Among different fertilizer treatments for pomato under study, maximum number of shoots per plant and prolonged harvest duration of tomato fruits were recorded at 75% RDF along with fertigation @ 4.56 g/m^{-2}. Least days to attain graftable shoots were found at 125% RDF. Minimum days (42.93) to first flowering in pomato plants and maximum plant height were recorded at 125% RDF + 2.28 g/m^{-2} fertigation. Yield attributes for potato and tomato in pomato plant were found at an upper hand in case of treatment T8 (100% RDF) + 6.84 g/m^{-2} (F) and T10 (75%).

11. **Performance of grafted tomato for horticultural and quality traits under polyhouse conditions**

- **Major findings:** This investigation was carried out for two successive years i.e. 2016-17 and 2017-18 to study the performance of grafted tomato with respect to growth, yield and quality of tomato under polyhouse conditions at Department of Vegetable Science and Floriculture, CSKHPKV, Palampur. Sixteen different rootstocks of tomato were grafted with scion cultivar 'GS-600'. The experiment was laid out in a Randomized Block Design with three replications and cleft grafting method was used. Plants grafted on rootstock LS-89 proved superior to other rootstocks for days to first flowering and days to first harvest. For prolonged harvest duration plants grafted on tomato rootstock Green Gourd found superior. Higher yield in terms of number of marketable fruits per plant, marketable fruit yield per plant and marketable fruit yield per square meter in were recorded plants grafted on tomato rootstock Green Gourd.

CHAPTER - 23

ROBOTIC VEGETABLE GRAFTING: NEW DIMENSION IN VEGETABLE GRAFTING

23.1 INTRODUCTION

Grafting technology is quickly evolving across the globe to meet the demands of expanding agricultural demands and diminishing land resources. Though grafting has several advantages, it is hampered by a number of issues, including graft incompatibility, which can arise as a result of graft union formation failure, which is highly reliant on environmental conditions, or a grafter's lack of ability. Vegetable grafting is a time-consuming and labor-intensive operation if carried out manually, hence efforts to mechanise grafting techniques have indeed been attempted for the previous two centuries. Due to issues such as low output and poor graft union formation, manual grafting is not as efficient as mechanised grafting. As a result, grafting robots are an important technique for the development of grafting plant production. With the aid of the "Grafting Robot," it is now feasible to graft crops mechanically. Grafting machines and robots are now available, which can speed up grafting, improve the survival rate of grafted plants, and make acquiring grafted seedlings more inexpensive.

The robotic grafting device can also eliminate vascular bundle mismatch during grafting, allowing grafted seedlings to recover faster and grow more vigorously. We can eliminate the need for a lot of manual work by employing robot technology. Large-scale agricultural production can be aided by mechanisation and automation in a short period of time. In order to minimise costs and enhance grafting success, further research should be focused on rootstock development, grafting robot efficiency, and the creation of acclimatisation facilities. Many propagators are interested in automated grafting or grafting robots because labour management is a major issue in large-scale propagation (more than 10 million plants per year) or propagation with seasonal needs.

23.2 SEEDLINGS SUITABLE FOR GRAFTING SYSTEM

The success or failure of grafting depends on the method of raising nursery plants. A high level of skill is required for the raising of nursery. The grafting system cannot always be applied to any kinds of seedlings. The methods of raising scions and rootstocks are very important to enhance the efficiency of grafting. For this purpose, seeds with a high germination rate are used. Healthy seedlings which are germinated in soil less media are suitable for grafting system.

23.3 HOW ROBOT WORKS?

The robot can easily detect the size and features of the vegetable plant to be grafted by pushing a button. Whenever the operator enters both a scion and a rootstock into the robotic machine, the robot carries, slices, and attaches the two parts of plants with the assistance of an elastic grafting clip. This entire procedure takes less than a second.

23.4 MECHANICAL DESIGN OF THE ROBOT

A robot is a programmable intelligent machine with different sensors and adjustable end-effectors that can do several materials handling jobs and is oriented and directed by a mechanism. Robots, in contrast to mechanisation duties, offer a variety of automation capabilities with flexibility, which may be extremely beneficial to plant production processes. The success rate of seedling cutting was ninety eight percent when the cutting knife rotation radius was 68 mm and the cutting speed was 120 r/min, according to the mechanism assessment and experimentation, which might fulfil the demand of grafting technique. The grafting rate was eight hundred eighty four plants per hour, with a 95.7 percent grafting success rate and a 96.8% survival ratio. As a result, grafting robots can tackle the issue of artificial grafting taking the place of physical labour.

Design of clamp feeding mechanism

To begin, organize seedlings in order using a bought grafting clip, then move the grafting clip to the push clamp position and clamp seedlings using the clamping device. An electromagnetic vibration disc, which consists of a vibration plate, an electromagnet, an armature, a spring component, a mounting seat, and a dampening rubber pad, is used to sort the grafting clip automatically. A controller's pulse current has the same natural frequency as the system and can stimulate an electromagnet. This arrangement is then brought into resonance, causing the armature and vibration plate to be rapidly drawn to the magnet.

The push and clamp mechanism's purpose is to push the grafting clip from its initial location to the point where the scion and rootstock are to be fitted while the clip is open. The clip is then closed to bite the rootstock and scion fitting pieces. The push and clamp mechanism is finally reset. In order to safeguard the grafted seedling, the clip should be moved gently. As a result, the mechanical device must fulfil the following criteria: The grafting clip stays open at its widest angle, forming the initial position to which the scion and rootstock are fitted; a buffer device is required to mitigate the impact generated by the grafting clip's closing action. The seedling anchoring system is required to retain the rootstock and scion pushed to an appropriate grafting location by the manipulator before being bitten by the grafting clip.

Mechanism for discharging grafted seedlings

A device to discharge grafted seedlings is required to transport grafted seedlings. As the transporting mechanism, a flat belt conveyor is used. Veins are required to increase the friction force between the roller and the belt core. The primary motor is a 24V DC stepping motor. Canvas is used for the belt's core. The power is transferred to the driving roller through the driving chain.

The device for discharging grafted seedlings)

23.5 GRAFTING ROBOTS

The grafting robot was created to alleviate the disadvantages of hand grafting. Grafting robots completely automate the grafting procedure. Some grafting robots are semi-automated, but new completely automated grafting robots have also been released. Robots have been developed that can speed up the procedure tenfold while maintaining the efficiency and precision of a traditional technique. Other robots, both semiautomatic and completely automated, are being developed in Spain, Taiwan, and Italy.

On the Japanese and Korean markets today, there are numerous varieties of basic grafting machines or completely autonomous robots that can successfully graft Solanaceous species. A specialist human can graft 150-180 seedlings per hour, but these machines and robots can graft 600-1200 seedlings per hour. The robots use computer programmes that allow them to pick, sort, and transplant homogeneous seedlings.

Semiautomatic grafting machine

A private Korean firm (Helper RoboTech) invented a multipurpose semi-automatic grafting machine, which many growers bought to graft tomato and pepper seedlings. For the first time, the Rural Development Administration (RDA) has created and marketed a robot that can graft both Cucurbitaceous and Solanaceous plants, such as cucumber and pepper and tomato. At the commencement of the first true leaf stage, scion seedlings are required. The genuine size of the leaf is 2-3mm. At the first genuine leaf stage, rootstock seedlings are required. Long hypocotyls (7-9 cm) are preferred.

The sort of graft done on the seedling is at side slits with a 45^0 cut in order to improve the contact surface between the two sections and therefore increase the likelihood of grafted seedling expansion. Because of its low pricing, varied features, and easy handling, this machine has been actively sold to many international nations in recent years. The robot was created in 2008, and 80 of them are already in operation in 13 different nations, including Italy, Mexico, the United States, China, Russia, Greece, Japan, and Spain.

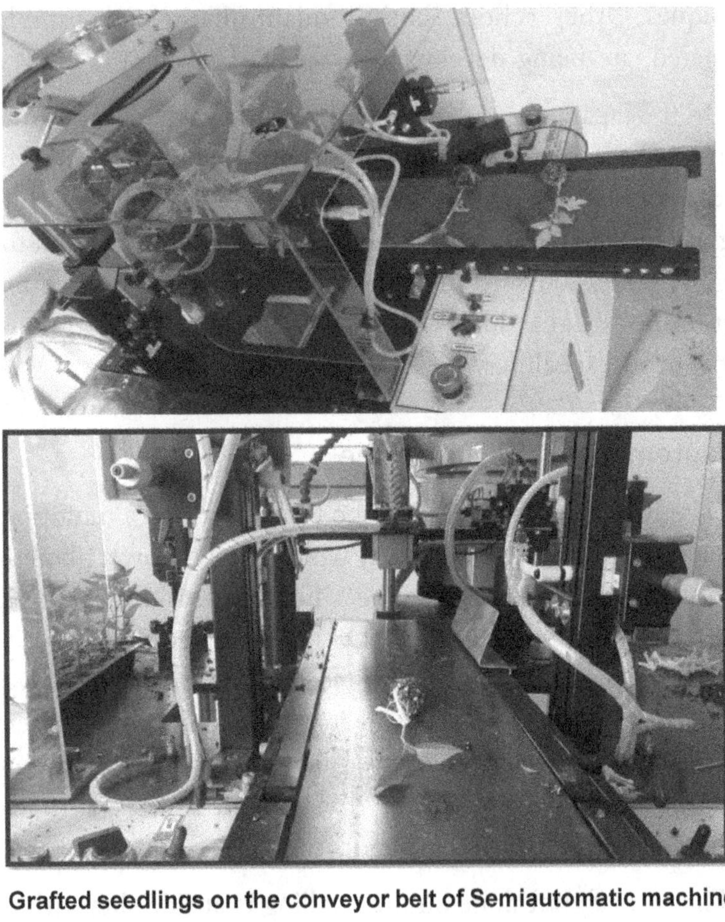

Grafted seedlings on the conveyor belt of Semiautomatic machine

Since the 1990s, a few semi-automatic grafting robots have been developed (Honami, Taira, Murase, Nishiura, & Yasukuri, 1992; Kubota, McClure, Kokalis-Burelle, Bausher, & Rosskopf, 2008; Lee et al., 2010; Masayuki, 1994), but these machines could only perform a limited amount of the operational plan to gain a comprehensive grafted seedling. The design of the automatic robotized machine for vegetable grafting presented is based on various primary features: ease of production, dependable operation, small size, modularity of components, operator safety, maximum sustainable productivity, and cheap cost.

- In Korea, the first semiautomatic grafting machine was created. Semi-automated grafting robots graft at a pace of 600-800 plants per hour (speed comparable of 5-6 skilled workers for cucumbers and 2-3 skilled workers for tomatoes) with a 95% success rate. Both cucurbits and tomatoes may be grafted by the model. A minimum of two people and one trained worker are required to help the robot. Someone must feed the plants to the machine in the semiautomatic unit. Workers feed seedlings with or without roots in this approach.
- Semiautomatic grafting machine from Japan (Iseki) that can only graft cucurbits and is sold in the Asian and European markets. It is capable of grafting 900 plants each hour with a 95% success rate. Robot need the assistance of two to three competent individuals.
- Semiautomatic grafting machine from Japan (Iseki) that can graft tomato and aubergine plants and is sold in Asia. The system can do 800 grafts each hour with a 95% success rate. Because the seedling size necessary for grafting is too huge by Japanese standards, the market is limited. The seedling size, on the other hand, is appropriate by American standards. The machine requires 2-3 persons to operate.

The Conic system (semi automated grafting machine) from Spain is capable of cutting tomato scion and rootstock at a certain angle. It has the ability to graft 400-600 plants every hour. This machine may be operated by only one person.

Fully automatic grafting machine

- Multiple seedlings are grafted at the same time in the completely mechanised procedure. For this technology, only one skilled person may load a tray of 100 or 200 plants into the machine, which then grafts the plants. Fully autonomous grafting robots often graft at a pace of 750 grafts per hour, with a 90% success rate. Seedlings are fed from trays on a belt conveyor by grasping hands in a completely automated system.
- In 1994, Yanmar Agricultural Equipment Co. (Osaka, Japan) produced the AG1000 robot for grafting Solanaceous plants, which was based on a 128-cell tray system and could graft an entire row of eight seedlings at once at a pace of 1200 grafts per hour with just one helping operator. Unfortunately, many of the grafting robots that were added to commercial propagators in the late 1990s and early 2000s are underutilised. Several concerns were raised by Kobayashi (2005), including the difficulty of reliably growing scion and rootstock seedlings that fulfil the size and quality standards demanded by grafting robots.
- In the Netherlands, completely automated equipment for grafting tomatoes was invented in 2006. This device, unlike semi-automated grafting machines, only requires one worker to provide rootstock and scion trays.
- The second-generation machine, which uses a grafting clip commonly used in the nursery business, began development in

2010 with enhanced operation and speed (more than 1000 grafts per hour). The devices are now being tested for large-scale grafting operations by the business.

New completely automated grafting robots have lately been created in Europe. The robots can achieve greater levels of performance, generating over 1000 grafted seedlings every hour with a success rate of up to 90% (ISO Group, 2013). These devices, however, are not ideal for ordinary Mediterranean nurseries due to their high cost and complexity. As a result, the process is still carried out manually in practically all circumstances. A national research project (PRIN 2009) was funded to create innovative equipment that is ideally suited for small & mid farms.

Layout of the machine

Two distinct trays, one for rootstocks and the other for scions, are put in the machine's intake station, near the two linear guides, as part of the overall arrangement. With each tray, the seedlings are removed from specialised manipulators and put on mobile jars controlled by two dedicated linear guides. The seedlings are transported to the vision station, where they will be classified and paired. The seedlings that are judged inappropriate are removed. A scion and a properly linked rootstock are then transported to the cut and junction station, where the grafting takes place. The grafting seedlings are then transferred to the last tray, where they will remain until they are fully developed.

The key benefit that this equipment aims to achieve is a safe, quick, and fully automated grafting procedure that allows the operator to utilise lower-quality pesticides and environmentally friendly treatments.

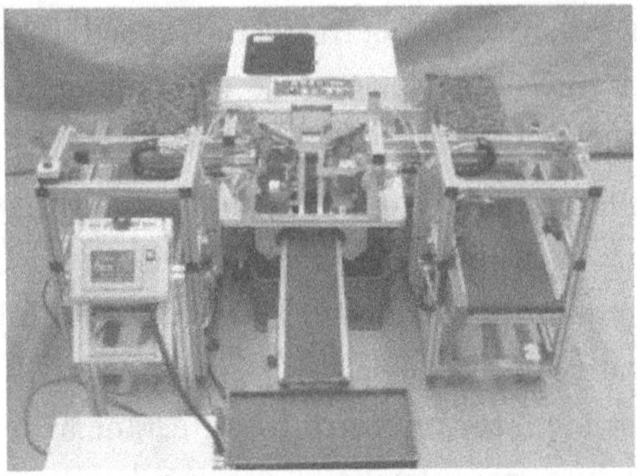

Fully automatic grafting machine

(**Source:** Kubota, CC., Mcclur, M.A., Kokalis - Burelle, N. and Bausher, M., 2008)

23.6 SUCCESS RATE OF MECHANICAL GRAFTING SYSTEM

The success rate of mechanical grafting is determined by the seedling quality and uniformity. The stem diameters of the rootstock and scion seedlings must match for effective grafting. When grafting seedlings, human operators utilise their 'eye-vision' to choose scion and rootstock seedlings with almost identical stem diameters. In an automatic grafting system, a camera can be used to grade and sort seedlings instead of human eyes. Any form of mechanical grafting technology has a success rate of 95% or above.

23.7 CONCLUSION AND FUTURE PROSPECTS

Grafting robots have shown to be an excellent investment on practically any scale, since they promote stronger food plants while also saving time and energy. This method may also be used to keep seedlings in good condition while being transported. Growers will soon be able to purchase seedlings that have been grafted by robots. Improvements in grafting robots and seedling production technology that may be automated are urgently needed, given the globalisation of grafting technology and rising demand from nations where labour management is a major concern in agriculture. Several research and development groups in Japan are actively working on improving the relatively old grafting robot technology or developing more flexible grafting robots. The requirement to meet increased demand while also lowering production costs necessitates the development of mechanisation for grafting. The introduction of superior rootstocks with various disease resistance as well as efficient grafting tools, including grafting robots, would considerably stimulate the global adoption of grafted vegetables.

Compressor- A major component of grafting Robot

Research scholar working on grafting robot at CSKHPKV-Palampur

REFERENCES

Anjali, Kumar, P., Sharma, P. and Thakur, V. 2020a. Influential exploration of rootstocks and grafting techniques on horticultural traits of parthenocarpic cucumber. *International Journal of Chemical Studies*, 8(1): 2839-2843.

Anjali, Kumar, P., Sharma, P., Kumar, S. and Thakur, V. 2020. Rootstock and grafting interaction studies for gummy stem blight (*Didymella bryoniae*) and horticultural traits in parthenocarpic cucumber. *International Journal of Current Microbiology and Applied Sciences*, 9(8): 3808-3817.

Anonymous, 2017-18. Indian Horticulture Database 2017-18. National Horticulture Board, Ministry of Agriculture, Government of India. Pp. 1-3.

Arvanitoyannis, I.S., Khah, E.M., Christakou, E.C. and Bletsos, F.A. (2005) Effect of grafting and modified atmosphere packaging on eggplant quality parameters during storage. *International Journal of Food Science and Technology*, 40: 311–322.

Bhatt, R.M., K.K. Upreti, M.H. Divya, S. Bhatt, C.B. Pavithra and A.T. Sadashiva. 2015. Interspecific grafting to enhance physiological resilience to flooding salinity. *Theoretical and Applied Genetics*, 128:667-679.

Bletsos, F.A. 2005. Use of grafting and calcium cyanamide as alternative to methyl bromide soil fumigation and their effects on growth, yield, quality and Fusarium wilt control in melon. *Journal of Phytopathology*, 153: 155-161.

Bletsos, F.A. 2006. Grafting and calcium cyanamide as alternative to methyl bromide for greenhouse eggplant production. *Scientia Horticulturae*, 107: 325-331.

Bletsos, F.A., C. Thanassoulopoulos, D. Roupakias. 2003. Effect of grafting on growth, yield and verticillium wilt of eggplant. *Hort Science*, 38: 183-186.

Chavez-Mendoza, C., Sanchez, E., Carvajal-Millan, E., Munoz-Marquaez, E. and Guevara- Aguilar, A. (2013). Characterization of the nutraceutical quality and antioxidant activity in bell pepper in response to grafting. *Molecules* 18: 15689-15703.

Ciccarese, F., Crino, P., Raccuia, S.A. and Temperini, A. (2012) Use of resistant cardoons as rootstocks for the control of verticillium wilt in globe artichoke. *Acta Horticulturae* 942: 201–206.

Cirulli, M., Bubici, G., Amenduni, M., Armengol, J., Berbegal, M., Jiménez-Gasco, M.M. and Jiménez-Díaz, R.M. (2010) Verticillium wilt: a threat to artichoke production. *Plant Disease Feature Article*, 94:1176–1187.

Colla G, Perez- Alfocea F and Schwarz D. 2017. Vegetable Grafting: Principles and Practices. *CABI, Cost European Cooperation in Science and Technology*, pp: 7-10.

Colla, G., F. Perez-Alfocea and D. Schwarz. 2017. Introduction to vegetable grafting. Vegetable grafting principles and practices. CABI, Pp.: 1-2.

Colla, G., Rouphael, Y., Cardarelli, M., Massa, D., Salerno, A. and Rea, E. (2006a) Yield, fruit quality and mineral composition of grafted melon plants grown under saline conditions. *Journal of Horticultural Science and Biotechnology*, 81:146–152.

Colla, G., Rouphael, Y., Jawad, R., Kumar, P., Rea, E. and Cardarelli, M. (2013) The effectiveness of grafting to improve NaCl and $CaCl_2$ tolerance in cucumber. *Scientia Horticulturae,* 64:380–391.

Colla, G., Y. Rouphael, C. Leonardi and Z. Bie. 2010. Role of grafting in vegetable crops grown under saline conditions. *Scientia Horticulturae*, 127: 147-155.

Colla, G., Y. Rouphael, C. Mirabelli and M. Cardarelli. 2011. Nitrogen- use efficiency traits of mini watermelon in response to grafting and nitrogen- fertilization doses. *Journal of Plant Nutrition and Soil Science*, 174:933-941.

Colla, G., Y. Rouphael, E. Rea and M. Cardarelli. 2012. Grafting cucumber plants enhance tolerance to sodium chloride and sulphate salinization. *Scientia Horticulturae*, 135: 177-185.

Colla, G., Y. Rouphael, R. Jawad, P.Kumar, E. Rea and M. Cardarelli. 2013. The effectiveness of grafting to improve Nacl and$CaCl_2$ tolerance in cucumber. *Scientia Horticulturae*, 164: 380-391.

Condurso, C., A.Verzera, G. Dima, G. Tripodi, P. Crino, A. Paratore and D. Romano. 2012. Effects of different rootstocks on aroma volatile compounds and carotenoid content of melon fruits. *Scientia Horticulturae*, 148:9-16.

Cushman, K.E. and Huan, J. (2008) Performance of four triploid watermelon cultivars grafted onto five rootstock genotypes: yield and fruit quality under commercial growing conditions. *Acta Horticulturae*, 782:335–337.

Davis, A.R. and Perkins-Veazie, P. (2005) Rootstock effects on plant vigor and watermelon fruit quality. *Cucurbit Genetics Cooperative Reports,* 28(29):39–42.

Davis, A.R., Perkins- Veazie, P., Sakata, Y., Lopez-Galarza, S., Maroto, J.V., Lee, S.G., Huh, Y.C., Sun, Z., Miguel, A., King, S.R., Cohen, R. and Lee, J.M. (2008). Cucurbit grafting. *Critical Reviews in Plant Sciences,* 27:50-74.

Delgado-Salinas, A. (1988) Variation, taxonomy, domestication, and germplasm potentialities in *Phaseolus coccineus.* In: Gepts, P. (ed.) *Genetic Resources of Phaseolus Beans.* Kluwer Academic Publishers, Dordrecht, The Netherlands, pp. 441–463.

Dor, E., B.A lperin, S. Wininger, B.Ben-Dor, V.S. Somvanshi, H. Koltai, Y. Kapulnik and J. Hershenhorn. 2010. Characterization of a novel tomato mutant resistant to the weedy parasites *Orbanche* and *Phelipanche* spp. Euphytica, 171: 371-380.

Edelstein, M., Burger, Y., Horev, C., Porat, A., Meir, A. and Cohen, R. (2004) Assessing the effect of genetic and anatomic variation of *Cucurbita* rootstocks on vigour, survival and yield of grafted melons. *Journal of Horticultural Science and Biotechnology,* 79: 370–374.

EL-Eslamboly A A. 2014. Effect of watermelon propagation by cuttings on vegetative growth, yield and fruit quality. *Egyptian Journal of Agricultural Research,* 92: 553-579

Fan, M., Z. Bie, A. Krumbein and D. Schwarz. 2011. Salinity stress in tomatoes can be alleviated by grafting and potassium depending on the rootstock and K-concentration employed. *Scientia Horticulturae,* 130:615-623.

Fuentes, I., S. Stegmann, H. Golczyk, D. Karcher and R. Bock. 2014. Horizontal genome transfer as an asexual path to the formation of new species, *Nature,* 511:232-235.

Ginoux, G. and Laterrot, H. 1991. Greffage de I' aubergine: reflexious sur le choix de porte-greffe. *PHM Revue Horticole*, 321:49-54.

Gisbert, C., Sanchez-Torres, P., Raigon M.D. and Nuez F. (2010). Phytophthora capsici resistance evaluation in pepper hybrids: agronomic performance and fruit quality of pepper grafted plants. *Journal of Food Agriculture and Environment*, 8:116-121.

Gisbert,C., J. Prohens, M.D. Raigon, J.R. Stommel and F. Nuez. 2011. Eggplant relatives as sources of variation for developing new rootstocks: effects of grafting on eggplant yield and fruit apparent quality and composition. *Scientia Horticulturae*, 128:14-22.

Giurca¢, D.M. (2009) Morphological and phenological differences between the two species of the *Phaseolus* genus (*Phaseolus vulgaris* and *Phaseolus coccineus*). *Cerceta¢ ri Agronomiceîn Moldova,* 42:39–45.

Goth, R.W., Haynes, K.G. and Barksdale, T.H. 1991. Improvement of levels of bacterial wilt resistance in eggplant through breeding. *Plant Disease*, 75(4):398-401.

Gruda, N. (2005) Impact of environmental factors on product quality of greenhouse vegetables for fresh consumption. *Critical Reviews in Plant Sciences,* 24, 227–247.

Guan, W. and Zhao X. 2014. Techniques for melon grafting. Document no. HS1257, Horticultural Sciences Department, University of Florida, Florida.

Hagerty, C.H., Cuesta-Marcos, A., Cregan, P.B., Song, Q., McClean, P., Noffsinger, S. and Myers, J.R. (2015) Mapping *Fusarium solani* and *Aphanomyces euteiches.* Root rot resistance and root architecture quantitative trait loci in common bean. *Crop Science,* 55:1969–1977.

Hazra, P., M.G. Som. 2015. History of vegetables and present status of vegetable cultivation in India.Vegetable science. Kalyani Publishers, New Delhi, Pp. 11-12.

Hernández-Bermejo, J.E. and León, J. (1992) Cultivos marginados, otra perspectiva de 1492. *Colleción FAO – Producción y Protección Vegetal,* 26:46–49.

Huang, Y., J. Li, B. Hua, Z. Liu. M. Fan, Z. Bie.2013b. Grafting onto different rootstocks as a means to improve watermelon tolerance to low potassium stress. *Scientia Horticulturae*, 149: 80-85.

Huang, Y., Kong, Q.S., Chen, F. and Bie, Z. 2015. The history, current status and future prospects of the vegetable grafting in China. *Acta Horticulturae* , 1086: 31-39.

Huang, Y., L.Q. Zhao, Q.S. Kong, F. Cheng, M.L. Niu, J.J. Xie, M.A. Nawaz and Z.L. Bie. 2016b.comprehensive mineral nutrition analysis of watermelon grafted onto two different rootstocks. *Horticultural Plant Journal,* 2: 105-113.

Huang, Y., Tang, R., Cao, Q. and Bie, Z. (2009) Improving the fruit yield and quality of cucumber by grafting onto the salt tolerant rootstock under NaCl stress. *Scientia Horticulturae,* 122, 26–31.

Huang, Y., Z. Bie, S. He, B. Hua, A. Zhen and Z. Liu. 2010. Improving cucumber tolerance to major nutrients induced salinity by grafting onto *Cucurbita ficifolia. Environmental and Experimental Botany*, 69: 32-38.

Huyskens-Keil, S. and Schreiner, M. (2003) Quality of fruits and vegetables. *Journal of Applied Botany and Food Quality*, 77:147–151.

Jang, Y., Huh, Y.C., Dong-Kum, P., Mun, B., Lee, S. and Um, Y. (2014) Greenhouse evaluation of melon rootstock resistance to Monosporascus root rot and vine decline as well as of yield and fruit quality in grafted 'Inodorus' melons. *Korean Society for Horticultural Science,* 32: 614–622.

Jang, Y., Moon, J.H., Lee, J.W., Lee, S.G., Kim, S.Y. and Chun, C. (2013). Effects of different rootstocks on fruit quality of grafted pepper (*Capsicum annuum* L.) *Korean Journal of Horticultural Science and Technology*, 31: 687-699.

Jasim AH and Abed HM. 2013. Effect of some treatments on rooting of cucumber cuttings (*Cucumis sativus* L.). *Euphrates Journal of Agriculture Science,* 5: 11-16.

Joeng, S.J. 1986. Effectsof several·rootstocks on the growth, fruit yield and quality of two cucumber cultivars. Institute of Food Development, Kyung Hee University, Suwon, Korea, Research Collection, 7:67-76.

Kachru S, Kuma P, Sharma P, Rana U,Upadhyay SK and Vats B ((2017): Effect of rooting hormones (IBAand NAA), Cow Urine and Growing Media on Growth,Yield and Quality of Tomato Plants propagated through stem cuttings. Indian *Horticulture Journal,* 7(3/4):193-195.

Kachru S, Kuma P, Sharma P, Rana U,Upadhyay SK and Vats B ((2017): Effect of indole butyric acid (IBA) , cow urine and growing media on root formation in tomato stem cuttings. *Himachal Journal of Agricultural Research,* 43(1):40-43.

Kalloo, G. 1993. Eggplant *Solanum melongena.* In: Kalloo, G., Bergh, B.O. (Eds.), Genetic Improvement of Vegetable Crops. Pergamon Press, Oxford, pp. 587-604.

Khah, E.M. (2011) Effect of grafting on growth, performance and yield of aubergine (*Solanum melongena* L.) in greenhouse and open-field. *International Journal of Plant Production,* 5:359–366.

Khah, E.M., Kakava, E., Mavromatis, A., Chachalis, D. and Goulas, C. (2006) Effect of grafting on growth and yield of tomato (*Lycopersicum esculentum* Mill) in greenhouse and openfield. *Journal of Applied Horticulture*, 8: 3–7.

King, S.R., Davis, A.R., Liu, W.G. and Levi, A. (2008). Grafting for disease resistance. *HortScience*, 43: 1673-1676.

Koren A, Klein E, J. Dieleman A, Janse J, Rouphael Y, Colla G and Mourão I (2017). Practical applications and specialty crops. *Vegetable Grafting Principle and Practices* Pp: 260-262.

Krumbein, A. and D. Schwarz. 2013. Grfating : a possibility to enhance health- promoting and flavour compounds in tomato of shaded plants? *Scientia Horticulturae*, 149: 97-107.

Kubota, C., 2016. Grafting manual: how to produce grafted vegetable plants. United state department of Agriculture, Pp: 1-2.www.vegetable grafting.org.

Kumar P, Sharma P, Raj D, Negi V, Rana S and Vats B. 2015. Growing capsicum, tomato and cucumber through stem cuttings. *Indian Horticulture* 60: 19-20.

Kumar P, L. Lucini, Y. Rouphael, M. Cardaarelli, R.M. Kalunke and G. Colla. 2015a. Insight into the role of grafting and arbuscular mycorrhiza on cadmium stress tolerance in tomato. *Frontier in Plant Science*, 6: 477.

Kumar P, Negi V., Sharma P., Raj D., Singh A. and Vats B. 2016. Effect of grafting on horticultural and quality traits in tomato scions grafted on potato rootstocks. *Indian Horticulture Journal*, 6(3): 352-354.

Kumar P., P. Sharma, B. vats and V. Sharma. 2018. Advances in vegetable grafting for mitigation of biotic and abiotic stresses in greenhouse vegetable production system. Techniques and sustainability of protected cultivation for Hi- valued vegetable crops, Pp. 129-162.

Kumar P., Sharma P. and Vats B. 2017. Effect of grafting tomato on different rootstock under protected environment in mid hills of NW Himalayan region. *Agric. Sci. Digest.*, 37(1):81-82.

Kumar, P., V. Negi, P. Sharma, D. Raj and S. Rana. 2015. Future wonder plant pomato. *Indian Horticulture*, 3-4.

Kumar, P., Y. Rouphael,M. Cardaarelli, and G. Colla.2015b.Effect of nickel and grafting combination on yield, fruit quality, anti-oxidative enzyme activities, lipid peroxidation and mineral composition of tomato. *Journal of Plant Nutrition and Soil Science*, 178: 848-860.

Kumar, S., Kumar, P., Sharma, P., Sankhyan, N. P. and Anjali. 2021. Effect of fertilization treatments on pomato growth and yield under protected environments. *International Journal of Current Microbiology and Applied Sciences*, 10 (2):1-8.

Kyriacou, M.C., G.A. Soteriou, Y. Rouphael, A.S. Siomos and D. Gerasopoulos. 2016. Configuration of watermelon fruit quality response to rootstock medicated harvest maturity and post-harvest storage. *Journal of the Science of Food and Agriculture*, 96: 2400-2409.

Labuda, H. (2010) Runner bean (*Phaseolus coccineus* L.) – biology and use. *Acta Scientiarum Polonorum – Hortorum Cultus* 9, 117–132.

Lee, J. M. 1994. Cultivation of grafted vegetables: Current status, grafting methods and benefits. *Hort. Science*, 29: 235-239.

Lee, J.M. and M. Oda. 2003. Grafting of herbaceous vegetable and ornamental crops. *Horticultural Review*, 28: 61-124.

Lee, J.M., C. Kubota, S.J. Tsao, Z. Bie, P.H. Echevarria, L. Morra and M. Oda. 2010. Current status of vegetable grafting; diffusion, grafting techniques, automation. *Scientia Horticulturae*, 127: 93-105.

Li, J.Y., H. X, G. X, J.J. Meng and Q.W. He. 2008. Higher chilling tolerance of grafted cucumber seedling leaves upon exposure to chilling stress. *Agricultural Sciences in China*, 7:570-576.

Lopez-Marin, J., A. Gonzalez, F. Perez-Alfocea, C. Egea-Gilabert and J.A. Fernandz.2013. Grafting is an efficient alternative to shading screens to alleviate thermal stress in greenhouse sweet pepper. *Scientia Horticulturae*, 149: 39-46.

Louws,F.J.,Rivard, C.L. and Kubota, C. 2010. Grafting fruiting vegetables to manage soil borne pathogens, foliar pathogens, arthropods and weeds. *Scientia Horticulturae*, 127: 127-146.

Lowman, M.S. and J.W. Kelly. 1946. The presence of hydriatic alkaloids in tomato fruit from scions grown on *Datura stramonium* rootstock. *Journal of American Society for Horticultural Science*, 48:249–259.

M. Amirullah, 2017. Effect of root stocks and grafting techniques on yield and quality of tomato (*Solanum lycopersicum* L.) under protected environment (M.Sc. Thesis).Chaudhary Sarwan Kumar Himachal Pradesh Krishi Vishvavidyalaya Palampur – 176 062 (H.P.) India.

Mišković, A., Ilin, Z. and Marković, V. (2008) Effect of different rootstock type on quality and yield of tomato fruits. *Acta Horticulturae,* 807: 619–622.

Mokhles, A., Kumar, P., Sharma, P., and Thakur, V. 2019. Effect of rootstock and grafting on quality of tomato (*Solanum lycopersicum* L.) under polyhouse conditions. *International Journal of Current Microbiology and Applied Sciences,* 8(7): 80-85.

Moncada, A., Miceli, A., Vetrano, F., Mineo, V., Planeta, D. and D'Anna, F. 2013. Effect of grafting on yield and quality of eggplant (*Solanum melongena* L.). *Scientia Horticulturae*, 149: 108- 114.

Mourao I., Moura M.L., L.M. Brito, J. Coutinho and S.R. Costa, 2016, unpublished results.

National Research Institute of Vegetables, Ornamental Plants and Tea. 2001. Current use and issues in grafted vegetable seedling production. Report No. 9, National Research Institute of Vegetable, Ornamental Plants and Tea, Japanese Ministry of Agriculture, Forestry and Fisheries.

Nawaz, M., Imtiaz, M., Kong, Q., Fei, C., Ahmed, W., Huang, Y. and Bie, Z. 2016. Grafting: a technique to modify ion accumulation in horticultural crops. *Frontier in Plant Science*, 7: 1457.

Negi, V., Kumar, P., Sharma, P., Raj, D., Singh, A. and Vats, B. 2016. Grafts compatibility studies in interspecific tomato-potato grafts. *Himachal Journal of Agricultural Research*, 42(1): 29-31.

Negi, V., Kumar, P., Sharma, P., Raj, D., Singh, A. and Vats, B. 2017. Horticultural and yield related traits as influenced by grafting tomato cultivars on potato rootstocks for higher returns. *Indian Journal of Ecology*, 44(2): 364-368.

Nilsen, E. and Orcutt D.M. 1996.The physiology of plants under stress-Abiotic factors. John Wiley and Sons, Inc., New York, 689.

Oda, M. 2002. Grafting of vegetable crops. Sci. Rep. Agr. & Biol. Sci. Osaka Pref. Univ. 54:49-72.

Oda, M.1999. Grafting of vegetables to improve greenhouse production. Food & Fertilizer Technology Center Extension Bulletin 480:1-11.

Panella, C., Nebauer, S.G., Quinones, A., Bautista, A.S., Lopez-Galarza, S. and Calatayud, A. 2015. Some rootstocks improve pepper tolerance to mild salinity through ionic regulation. *Plant Science*, 230: 12-22.

Peri, C. (2004) Qualità Nelle Aziende e Nelle Filiere Agroalimentari. Hoepli, Milano, Italy.

Qun Sun, Ying Zhao, Anfu Guo, Dongjie Zhao and Chong Wang. 2016. Design and Test of a Solanaceae Grafting Robot. *The Open Mechanical Engineering Journal*, 10: 1-12.

Ram, H.H. 2017. Indian national vegetable improvement programme. Vegetable breeding principles and practices. Vegetable science. Kalyani Publishers, New Delhi, Pp. 18.

Rouphael, Y., Cardarelli, M., Lucini, L., Rea, E. and Colla, G. (2012) Nutrient solution concentration affects growth, mineral composition, phenolic acids and flavonoids in leaves of artichoke and cardoon. *HortScience* 47, 1424–1429.

Rouphael, Y., Schwarz, D., Krumbein, A. and Colla, G. 2010. Impact of grafting on product quality of fruit vegetables. *Scientia Horticulturae*, 127: 172-179.

S.B. Tian, J.F. Yang, R.L. Wang, D. Xu, and T. Li. 2014. "Optimiza-tion experiment of operating parameters on vibration sorting-clip device for vegetable grafting machine". *Transactions of the Chinese Society for Agricultural Engineering*, 30: 9-16.

Sakata, Y., Ohara, T. and Sugiyama, M. 2008. The history of melon and cucumber grafting in Japan. *Acta Horticulturae*, 767: 217-228.

Sánchez-Rodríguez, E., Ruiz, J.M., Ferreres, F. and Moreno, D.A. (2012) Phenolic profiles of cherry tomatoes as influenced by hydric stress and rootstock technique. *Food Chemistry*,134: 775–782.

Sarswat, S., Kumar, P., Sharma, P. and Thakur, V. 2020. Standardization of robotic grafting in bell pepper (*Capsicum annuum* L. var. *grossum* Sendt.). *International Journal of Current Microbiology and Applied Sciences*, 9(3): 1410-1418.

Schneider, K.A., Grafton, K.F. and Kelly, J.D. (2001) QTL analysis of resistance to Fusarium root rot in bean. *Crop Science* 41, 535–542.

Schwarz, D., Rouphael, Y., Colla, G. and Venema, J.H. 2010. Grafting as a tool to improve tolerance of vegetable to abiotic stresses: thermal stress, water stress and organic pollutants. *Scientia Horticulturae*, 127: 162-171.

Shewfelt, R. (1999) What is quality? *Postharvest Biology and Technology* 15, 197–200.

Shibuya, T., Itagaki, K., Wang, Y. and Endo, R. 2015. Grafting transiently suppresses development of powdery mildew colonies, probably through a quantitative change in water relations of the host cucumber scions during graft healing. *Scientia Horticulturae*, 192: 197-199.

Soteriou, G.A., Kyriacou, M.C., Siomos, A.S. and Gerasopoulos, D. (2015) Rootstock-mediated effects on watermelon ripening behavior and fruit physicochemical and phytochemical composition. *Acta Horticulturae*, 1079: 707–714.

Tateishi, K. 1927. Grafting watermelon onto pumpkin. *Japanese Journal of Horticulture (Nihon-Engei Zasshi)*, 39:5-8.

Temperini, O., Calabrese, N., Temperini, A., Rouphael, Y., Tesi, R., Lenzi, A., Carito, A. and Colla, G. 2013. Grafting artichoke onto cardoon rootstocks: grafting compatibility, yield and Verticillium wilt incidence. *Scientia Horticulturae*, 149: 22-27.

Temperini, O., Calabrese, N., Temperini, A., Rouphael, Y., Tesi, R., Lenzi, A., Carito, A. and Colla, G. (2013) Grafting artichoke onto cardoon rootstocks: graft compatibility, yield and Verticillium wilt incidence. *Scientia Horticulturae* 149, 4, 22–27.

Trinchera, A., Pandozy, G., Rinaldi, S., Crino, P, Temperini, O. and Rea, E. (2013a) Graft union formation in artichoke grafting onto wild and cultivated cardoon: an anatomical study. *Journal of Plant Physiology* 170, 1569–1578.

Trinchera, A., Pandozy, G., Temperini, A., Caprio, F., Rea, E. and Rinaldi, S. (2013b) Grafting of globe artichoke plants onto wild and cultivated cardoon: agronomical and physiological aspects. *Acta Horticulturae* 983, 231–235.

Tirupathamma T. Lakshmi, C. Venkata Ramana, L. Naram Naidu and K. Sasikala 2019. *Vegetable grafting: A multiple crop improvement methodology*, 33(3): 1-10.

Tsaballa, A, Athanasiadis, C., Pasentsis, K., Ganopoulos, I., Nianiou-Obeidat, I. and Tsaftaris, A. 2013. Molecular studies of inheritance grafting induced changes in pepper (*Capsicum annum*) fruit shape. *Scientia Horticulturae*, 149: 2-8.

Ulukapi, K. and Onus, A.N. (2005) Effect of grafted tomato (cv. 191 F1) on yield and fruit quality. In: GAP IVth Agricultural Congress, Sanliurfa, Turkey, pp. 1314–1317.

Verzera, A., Dima, G., Tripodi, G., Condurso, C., Crinò, P. and Romano, D. (2014) Aroma and sensory quality of honeydew melon fruits (*Cucumis melo* L. subsp. *melo* var. *inodorus* H. *Jacq.*) in relation to different rootstocks. *Scientia Horticulturae*, 169: 118–124.

Vinkovic-Vrcek, I., Samobor, V., Bojic, M., Medic-Saric, M., Vukobratovic, M., Erhatic, R., Horvat, D. and Matotan, Z. (2011) The effect of grafting on the antioxidant properties of tomato (Solanum lycopersicum L.). *Spanish Journal of Agricultural Research* 9, 844–851.

Yamakawa. K. 1982. Use of rootstocks in Solanaceous fruit-vegetable production in Japan. *Japan Agricultural Research Quarterly, Ibaraki.* 15(3): 175-180.

Yassin H. and Hussen S.(2015) global Journal of Science Frontier Research: D Agriculture and Veterinary,15 (1).

Yetisir, H. and Sari, N. (2003) Effect of different rootstock on plant growth, yield and quality of watermelon. *Australian Journal of Experimental Agriculture*, 43:1269–1274.